A Lively Hive

By Alex Tuchman

A BIODYNAMIC BEEKEEPING GUIDE FOR HONEYBEE HEALTH

All interior photos by Alex Tuchman, Anthea van Geloven, Helena Kiko-Cozy and Vivian Struve-Hauk.

Copy editing by Carol Frazier Johnson
Cover design by Emily Kitching
Cover photo by Alex Tuchman

www.thehometeadpress.com

Table of Contents

Foreword

"Anything will give up its secrets
if you love it enough."

— *George Washington Carver*

Dear reader,

It is with profound inner satisfaction that I see this manual being published. This beekeeping method, derived from a deeper understanding of the honeybees' nature and own needs, has been practiced and promoted by me for more than 40 years. Its success has been instrumental in helping to bring about the changes we witness now. For a good 20 years, a growing number of attempts have emerged, challenging beekeeping methods that, for about a hundred-fifty years, had one goal and one goal only: to optimize honey harvest and make beekeeping easier. Invention after invention furthered this goal. And there was great pride in such achievements!

What helped to actually bring about a change were a number of crises: the arrival of the varroa mites in the USA in the late 1980s, resulting in tremendous losses of colonies in the 90s, and ten years later "colony collapse disorder," witnessing even more losses. On the positive side, these crises were accompanied by a growing scientific awareness and warnings that the unhindered exploitation of nature and her beings would eventually result in a planet that could not sustain life.

It became clear at the outset of the new millennium that a radical change in our attitude and goal of working with domesticated animals—one aspect of the greater challenge—was needed.

"What on Earth are we doing?" was an ethical stirring of conscience for a growing number of individuals. "Grass-fed beef" and "cage-free chicken" became signs not only for a call for better nutrition but also for a moral awakening. In beekeeping, "organic," "natural," "treatment-free,"and "sustainable" beekeeping voices filtered into the existing paradigm which was facing a growing number of existential crises.

It had become clear that the loud call "to make bees more profitable" needed to be balanced by the quest to understand this unusually complex and utterly magnificent insect, the honeybee, at a deeper level, looking at what she herself needed in order to be healthy and regain her previous vitality.

Spikenard Farm has spearheaded such a change in attitude and methodology for a good number of years—with exemplary results.

It is this overdue and ongoing change that lets us breathe a sigh of relief, encouraging us to become humble stewards of the creatures that are entrusted to our care. It is this new path that gives hope by healing the wounds that a shortsighted, profit-only-oriented goal has smitten. Leaving the old, obsolete path is not easy for everyone, but whatever is not achieved freely by a change in attitude and intention and by re-directing the needle of our moral compass eventually will be forced into change by the catastrophes that loom on the horizon, witnessed by all whose eyes are wide open.

I can only encourage new and old beekeepers to try the path that is described in such eloquent and yet precise language by Alex whose devotion to this important work I so deeply appreciate. The path described in this book, from spiritual insights to "nuts and bolts," will certainly deepen for you the love and understanding the honeybee deserves and will become a source of inspiration for your personal relationship to this magnificent insect.

The rewards will be bountiful, not only in respect to beekeeping.

Gunther Hauk
March 2021

Introduction

We are in the midst of a beekeeping revival—a distinct period of change wherein a new age of holistic beekeeping methods and practices are gaining firm ground due especially to their success in maintaining healthy colony survival rates in contrast to the conventional/professional beekeeping methods that are failing to come up to speed with the changing times.

The "conventional methods," however, have provided a great service to beekeeping in all of the tools and technologies that have aided the beekeeper in coming into a more scientific, sense-based, and physically intimate relationship with the life of the honeybee colony. With the invention of a new hive type by Reverend Lorenzo L. Langstroth in 1850 (now known as the Langstroth Hive), we were finally able to explore the inner life of honeybees in ways that we had never seen before. Unfortunately, this new hive body has been altered more and more by those who are focused on modes of production above all else (we certainly cannot blame Langstroth for this!). The one-sided, economically driven relationship between bees and beekeeper has now held sway for over a century, following the course of modern humanity's exploitation of the natural world in all its facets. While this mindset may persist for a few decades yet, it is clear to anyone who has a sense for human worthiness and truly looks at the nature of this relationship that it cannot be sustained. A re-awakened moral sensibility for nature is absolutely necessary to rescue our Earth—and the bees. The beekeeping revival that is just beginning to enter into the public conversation is given force by individuals who are on this journey towards healing the relationship between the human being and the natural world.

Today, keeping our bees alive is becoming the foremost measure of successful practices. In 2015—2020 alone, the USA lost on average over 45% of its colonies each year. And commercial beekeepers, too, are going bankrupt, getting out of business, and retiring in numbers that are just as shocking as their hive losses. Now that the methods that exploit the bees are going bankrupt, methods that truly care for the bees are becoming more recognized for their lasting value.

As more beekeepers and hobbyists are waking up to the crisis, there has opened up a great creative space, as the focus has gone from "What can I get out of the bees?" to "How can we do this correctly and in line with the instincts and health of the colony?" The influx of new ideas paired with old wisdom brings the benefit of a full overhaul—a re-evaluation of where we are now, what is working, and what is not. It allows beekeepers, especially "hobbyists" (or backyard beekeepers, 1-10 hives) and "sideliners" (10-200 hives, part time) to build on the successes of the past while applying new thoughts to the ancient practice of keeping bees.

My goal in this work is to elaborate in detail on biodynamic beekeeping practices, many of which were first passed to me by my teacher, co-worker, and friend, Gunther Hauk. These methods are derived from the indications given by the contemporary scientist, poet, philosopher, and spiritual scientist, Rudolf Steiner, who is the founder of anthroposophy, and the biodynamic method in both agriculture and beekeeping. As a gardening teacher at the world's premier Waldorf School, Hauk learned beekeeping methods in the world's oldest biodynamically managed educational garden from Klaus Matzke, who learned from his father-in-law, Harald Kabish, who was one of the first regional biodynamic consultants in the early years after Rudolf Steiner's famous Agriculture Course, given in 1924.

It was in 1996 that Hauk gave his first "organic" beekeeping workshop in the United States in response to the New York Times article "The Hush of the Hives." At that time, Hauk had been beekeeping by the biodynamic method for 20 years and knew that the writing was on the wall–he could see that beekeeping was headed towards a major collapse. When the first news about the varroa mite reached America, and then later, when news of colony collapse disorder (CCD) shocked the world, he knew that he had the wherewithal to help address the mounting challenges that beekeepers were facing. In 2007, Hauk and his wife Vivian Struve-Hauk founded a non-profit organization, Spikenard Farm Honeybee Sanctuary, in order to give the practical grounds for the application of these methods, further the research into biodynamic beekeeping, and serve the public as an educational institution whose continued mission is to inspire and teach beekeeping practices that honor the needs of the honeybee above all else.

This book is written in honor of Hauk's pioneering work in this field in the United States, and it is meant as a direct accompaniment to his own book, *Towards Saving the Honeybee*, which was first published in 1996, republished in 2017. His book goes in-depth into the philosophies of the biodynamic method which consist of:

- Hive forms that respect and approach ROUNDNESS, and equipment that promotes HIVE WARMTH and SCENT.
- The bees make their own NATURAL WAX COMB.
- The hive is respected as an ORGANISM and nurtured in its inherent tendencies for the creation of the right amount of workers, drones, and queens.
- The celebration of SWARMING as the most vital and sustainable method of expanding the apiary, breeding queens, and selling bees/hives.
- The adoption of a conservative approach to the FEEDING of sugar and HARVESTING of honey.

• A holistic approach to HEALING and TREATMENTS.

• An understanding of the land and the bees as one, and a focus on providing the best BEE FORAGE.

These are the requirements that the honeybees need to sustain a healthy colony. These truths gave rise to beekeeping methods that honor the needs of the honeybees without taking advantage of the bees to the detriment of their health.

Hauk has always given people the gift of freedom in his lectures and indications on beekeeping methods. We have been working together, keeping bees, and teaching together since 2014, and I have yet to see him give a student a recipe or a direct "you must do it this way or that." His method is incredibly encouraging and has left his students and himself in freedom to continue to learn and grow—to change with the times and with the bees. It's awe-inspiring to observe Hauk, who after 40 years of beekeeping is always learning something new.

The fact that Hauk did not write the book that you are now reading is due to these virtues. *Towards Saving the Honeybee* is all one would need if one's own practical path could be creatively and courageously built from the ideals presented therein. The success of beekeeping ultimately depends on the personal relationship between bees and beekeeper. *Towards Saving the Honeybee* strives for a depth of understanding of the honeybee—an understanding that can give rise to creative solutions and methods of beekeeping that can help guide humanity out of this current crisis. Paraphrasing Hauk, we learn that there is no true love of the bees without the work and earnest striving that is necessary to understand what they need. If we can place ourselves in this relationship to the bees, we can continue unfolding our relationship and understand more deeply the wisdom of the bees and what we can do to provide for their health.

As a necessary and long-requested addition to *Towards Saving the Honeybee*, this book will provide recipes and detailed guidance for beekeepers, and it is intended to serve the needs of those who are yearning for practical indications in bee-keeping. The risk of "fixing" these methods in a written work has been taken by the author consciously, as the importance of having this information available has become a more pressing need. Within the conversation and context of contemporary beekeeping, the voice of biodynamic beekeeping must become available to a wider audience and stand in service to beginners as well as to seasoned practitioners who are yearning for another way forward and a strong, clear, and comprehensive basis for relating to biodynamic beekeeping. My goal in writing these methods on paper was to bring the ideals and spiritual truths down to Earth for others to try out and to work with and to have as a starting guide for the development of their own practices.

All of the methods presented herein are practiced and taught at Spikenard Farm Honeybee Sanctuary, located in Floyd, Virginia. No book can replace the first-hand, experiential learning that is gained through attending a workshop at Spikenard and taking part in the community of striving beekeepers who join us here with the bees every year. Spikenard is intended to serve not only as a well-spring of information and practical guidance but also as a transformative community of individuals who are working together in mutual support to affect the changes that are necessary for our time. This is a community that can only grow rightly into the future as a culture of the heart, a culture of love. In this striving, we are consciously following the wisdom path of the honeybees. We welcome you to join us in this journey!

Alex Tuchman
January 2021
Spikenard Farm Honeybee Sanctuary
Floyd, Virginia

Chapter 1

Setting the Context
for Beekeeping

The aim of our relationship is the well-being of both the honeybees and the beekeeper. Health-engendering practices are no longer a matter of course within the dominant beekeeping culture, just as the environment surrounding the bees is often no longer a given as a healthy basis for their life. The journey forward with the bees demands that our growing understanding of life-giving beekeeping methods be paired with actions to nurture the environment which surrounds them.

1

THE ROLE OF A BEEKEEPER

Before we dive into what we as beekeepers must practically do, I encourage you to consider the questions:

What are my beekeeping goals?
Why do I want to keep bees?

Do I want to have honey to eat, sell, or share;
to have wax for making candles, salve, or other products;
to have propolis for medicinal tinctures;
to have better pollination of the orchard, farm, garden, or crops;
to break-even with my investment;
to help the bees survive;
to contribute to the bees' health and vitality;
to deepen my relationship with the natural world;
to keep bees as a profession;
to start a new hobby;
to follow a calling….

What can you add to this list? It is likely that a combination of goals and interests are living in you. I encourage you to clarify these goals at the outset and revisit them as you go forward as part of the learning process.

These basic questions lead us to the first steps of being a beekeeper (or of reinventing yourself in your beekeeping practice!) because how much effort you are willing to put forth, how much time you have, how you use the resources available to you, the type of hive you will choose, and the type of beekeeping training you seek out are all aspects that must be navigated depending on the directions you choose in this initial goal-setting exercise.

The common saying, "There are as many ways of keeping bees as there are beekeepers" has truth to it. In our contemporary beekeeping culture, you will find strong opinions in excess. This can be difficult to confront, especially if you have ideals such as the ones offered in this book, which are based in a different mindset and founded with a different intention than the conventional counterparts. In approaching conventional beekeeping, you can expect that much of what you hear might not feel right or make logical sense in your mind (i.e. the common suggestion of feeding the bees sugar because feeding the bees honey is taught to be dangerous). However, an argument over ideals will not be well-substantiated on your part and not likely be respected unless you have first-hand experience of your own. No one can tell you that you are wrong if you are able to claim, "I have experience with this. I do it this way, and it works for me." If you are reading this book and feel good about the attitude and approach presented so far, rest assured, knowing that you are not alone but are joined by many other beekeepers who are swimming against the main-stream currents.

We can set up systems of beekeeping with our goals in mind, but we also may have to sacrifice one intended benefit or another, depending on the year and depending on the hive. The bees are great teachers of humility and gratitude, and it is recommended to approach your role as a beekeeper with a willingness to regularly practice both. The gifts that we beekeepers then receive from healthy hives in productive years can be celebrated with joy and thankfulness.

A simple but incredibly important question arises in relation to the beekeeper's desire to serve the bees:

What do the bees need me to do in my role as a beekeeper, and when do I leave the bees to do what they know how to do best?

This question is held through every step of this beekeeping manual. In our work at Spikenard Farm Honeybee Sanctuary, we seek not to place the bees and the beekeeper opposed to each other, but to keep them in a mutually beneficial relationship. The role of the beekeeper in this approach is concerned with engaging in activity that will benefit the bees in their health and vitality.

We know from experience that a colony can be kept successfully—with longevity, health, and with full opportunity to express the wholeness of its being—in many different hive types, each of which require different methods and practices for success. Further, we know that from healthy bees we can receive gifts of the most healing substances: honey, wax, and propolis, and gather them harmoniously without detriment to the hive itself. We also know that all beekeepers have their own gifts and challenges, and will make their own choices— we wish to leave you free to decide how to engage in practices that make logical sense, feel right in your heart, and which you can carry out yourself.

It is in service of a flexible and actively growing universal approach to beekeeping that we have been working to develop the Spikenard Method of biodynamic beekeeping which is presented in this book.

The Spikenard Method is a *path* of knowledge at its foundation. In order to answer the question of when to act versus when to hold back, we ask what the bee needs and then what we can do to serve those needs. We strive for knowledge of her basic life processes, instincts, and wisdom, and approach the mysteries and enigmas of her being with awe, reverence, and devoted attention. We ask what the bees need and then learn to listen. In this way, we begin to develop within ourselves the most practical tool in beekeeping: love. If we approach the bees with love and allow it to continually grow between us, new life will be infused into our beekeeping practice day-by-day. The bees will be comfortable with our presence and will be encouraged by previous positive experiences when we come to care for them again with understanding.

The path of knowledge that we work from has its compass needle pointed towards the development of human love. The wisdom of the honeybees is the compass needle itself.

"That which we experience only at a time when our hearts develop love is the same thing that is present as a substance in the beehive. The whole beehive is permeated with life based on love."
– Rudolf Steiner, **Bees** *1923*

LANDSCAPE CONSIDERATIONS

Hive Placement:

Hives should be oriented South-East. East—for the morning sun to get them out early, and South—for the most warmth and light in the Winter (in the Northern Hemisphere.)

Windbreaks:

It is the wind (draft), rather than the cold temperatures, that has the greatest negative impact on the bees, especially for overwintering. Windbreaks are very important!

Height Off the Ground:

Keeping hives off the ground reduces easy-access for unwanted visitors such as ants, mice, and skunks. We recommend a 1.5' stand.

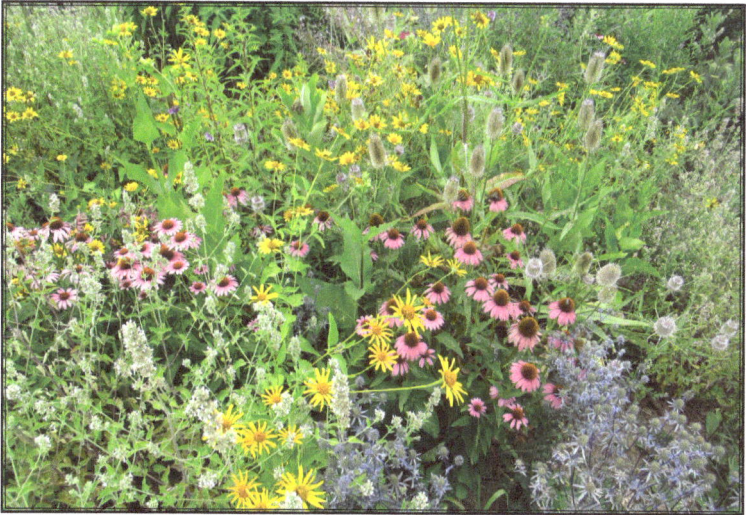

Growing Clean Bee Forage:

What can you plant for your bees? Offering healthy forage is one of the best ways to enhance the value of the honey and pollen and provide the bees with a diverse medicine cabinet!

What Wildflowers are Growing?

And how can you promote them? Wildflowers in ditches and roadsides, along with neighboring lands and forests are incredibly important resources to evaluate. Worker bees generally forage within a three-mile radius of their hive (depending on the ecosystem and forage density). Bee-keeping clubs and associations have some of the best local knowledge for natural sources of nectar and pollen. See also the NASA publication: https://honeybeenet.gsfc.nasa.gov/Honeybees/Forage.htm.

How Do You Mow the Lawn?

Simply raising the mowing bar two inches higher can provide a huge source of forage for the bees, as pictured here with the white Dutch clover and plantain, where we often get 4 blooms per year by adjusting our mowing techniques.

We let the first bloom go until 2/3 of the flowers have senesced before mowing again on the 4'' setting of our riding mower. Wait and repeat!

For better or for worse, the bees naturally bring us beyond our fence-line and into relationship with the land-management practices of our neighbors. The bees are sensitive creatures, vulnerable to the activities spread across the landscape in the bees' range of foraging, which is a variable range depending on your bioregion, but generally extends out from the beehive in a three-mile radius. For some people, this will present far more challenges than for others. It is recommended to go with honey and helpful information about the bees and native pollinators* to your neighbor if you need to have a conversation about spraying toxic chemicals.

*In our experience, the native pollinators and honeybees live in symbiosis in the environment. Honeybees enhance the quantity of available forage year-by-year through their abundance and ecosystem services, which benefit the whole. In this way, honeybees could be considered a keystone species. A lengthy presentation of native pollinators and their greater roles in ecosystems is not within the scope of this book but is noted here as an essential and important part of landscape considerations.

BEEKEEPING TOOLS AND EQUIPMENT

If you're getting started with beekeeping for the first time, or beginning again, we recommend starting with two Langstroth Hives. To learn and practice keeping bees in Langstroth Hives provides a solid basis from which to pursue other creative hive-forms, such as the Sun Hive, the Warre Hive, the Top Bar Hive, the Layens Hive, etc.

Langstroth Hives have the advantage of giving the beekeeper the most hands-on experience with the bees, helping to train the eye, the hands, and the heart all together. For example, it is a great help to know what a healthy hive looks like by inspecting the combs in the file-cabinet-like system that the Langstroth Hive provides. To go a step further, we can pair what a healthy hive looks like with what a healthy hive smells and sounds like. In this way, we can learn over time to experience hive health by smell and sound in a way that can help us relate to a more hands-off hive type, such as a Sun Hive, without opening it. Of course, the same is true for learning what a sick, diseased, or infested hive smells like, or the sound of a queen-less hive. (More on this in the Developing our Sense Perception section see page 62.)

Aside from the normal health checks and fertility support that are part of beekeeping in any hive type, the main work of beekeeping in a Langstroth Hive consists of expanding hives as they grow in the summer and tightening up the hive space for the winter. For the beginner, you have a steep learning curve and a lot of information you need to gather before you can finally be successful.

Langstroth Hives tend to be the easiest to build from a kit and the cheapest to purchase, and they are the most popular hives in the world, making access to information, mentors, and extra supplies accessible on every continent.

Note: I am sure that there will be many in the natural and biodynamic beekeeping movement who will disagree with the choice to guide beginners towards beekeeping in Langstroth Hives. I have consciously taken this choice because I felt the information would be helpful for the most people and have the widest possible impact on common beekeeping practices, in addition to the reasons mentioned above. Further elaborations on working with other hive shapes will be possible in a future book; however, the beekeeping practices outlined in this manual, to both a great and lesser degree, will be able to be applied and adapted by the reader to other hives, in most cases.

To get started with a Langstroth Hive and work with it by the Spikenard Method, here is a list of what you need to buy (this is not a conventional list—it excludes conventionally common, but unnecessary, equipment such as queen excluders, plastic foundation, sugar feeding trays, etc.)

Equipment You Need To Start:

Langstroth

Companies in the USA to purchase beekeeping supplies: Betterbee, Brushy Mountain, Dadant, Rossman Apiaries, Mann Lake, Walter T. Kelly. Plans for building bee equipment: www.beesource.com

2	Eastern Pine deep hive body	43.90
2	Eastern Pine medium super	36.30
20	Deep frames (wedged/grooved)	27.90
20	Medium frames (wedged/grooved)	27.90
1	Unwired brood foundation (synthetic wax)	43.00
1	Wooden inner cover	11.50
1	Galvanized outer cover	19.95
1	Pine reversible bottom board	18.85
1	Screened bottom board	21.95
1	Debris tray	4.95
1	J hook hive tool/hook end hive tool	10.95
1	Wooden entrance closure	1.55
1	Standard smoker	32.95
1	Bee brush	5.95
1	Ventilated helmet	19.90
1	Premium tie-down veil(round)	17.75
	Total	**$ 345.25**

Your hive set up starts with a pine reversible bottom board.

The debris tray sits in the bottom board.

Screen bottom board, sometimes called a varroa screen.

The screen bottom board sits on top of the solid bottom board, with the debris tray inserted into the slot that is made between the two. Note: the solid bottom board is facing backward, and the screen bottom board is facing forward. This creates the debris tray slot at the back of the hive.

Pictured from the back of the hive. The slot needs to be closed with a piece of wood, which we call a "back closure."

Put in the back closure so that the slot can be closed, unless you are checking the debris tray.

Eastern pine deep hive body, often called a "deep."

Eastern pine medium super, often called a "super."

Each box holds 10 frames. A deep frame is pictured on the left, a super frame on the right.

The inner cover is placed above the highest box.

Hive set up with inner cover.

Entrance reducers make the entrance smaller and easier for the hive to control temperature, airflow, and activity.

Hive with entrance reducer.

The telescoping outer cover goes above the inner cover and forms the roof of the hive set up.

Whole Langstroth hive set up from the front.

Hive tools: the J-hook hive tool on the right and the standard hive tool on the left.

Hive tools are used to loosen the frames and pry them out during hive inspections.

Bee brushes are used to gently guide the bees to where you want them to walk. This helps avoid squishing bees!

Bee brushes can be used to guide a swarm into a new hive body.

The woodenware that makes up this Langstroth Hive can also be referred to more generally as the hive body. A hive body can be a hollow tree, the soffit in your house, the barn wall—it creates the empty cavity that the bees live in and utilize as their outer boundary. When human beings domesticated the honeybee, the beekeepers first created hive bodies that would allow them to keep the bees close to home and make harvesting possible. Round logs, skeps, baskets, and clay pipes were used for thousands of years with minimal physical intervention—mostly honey and wax harvest. Humanity's recent creativity and exploration into new hive types and different hive forms have produced at least a dozen new hive bodies in the last century that beekeepers all around the world are using—each with a slightly different goal, ideal, or human desire behind its creation.

At the Sanctuary, we have seven different hive types that we use with success, and we continue to explore the creation of new, and correction of established, hive bodies with an eye to serve the honeybees. We do not advocate one hive body to be better than another. They all have advantages and disadvantages which point directly to the relationship between the bees and the beekeeper. At present, the common discourse between beekeeper and bees places them in opposition, i.e., if one aspect of the hive construction is helpful for the beekeeper, it is usually to the detriment of the bees, and vice versa. This way of looking at things obscures the importance of our task as beekeepers, which is not to put our wishes and desires in opposition to those of the bees but to learn to give to them all that they need to be healthy and thrive. Our experience at Spikenard Farm Honeybee Sanctuary shows that honeybees can thrive in all of the different hive shapes. It is developing a true understanding of the honeybee and her needs that elevates our relationship and lends itself to successful beekeeping methods that allow the bees to be strong, productive, and vital.

THE DEBRIS TRAY – TECHNOLOGY TO ENNOBLE THE LANGSTROTH

Most of the inventions in the last 100 years of beekeeping do not have the bees' health in mind. But there is one that was invented in response to the rising challenges that came with the varroa mite that is absolutely worth all the rest combined. This is the debris tray.

This monitoring tool is one of the central factors in our ability to successfully guide Langstroth Hives through the season. It sits on the bottom board of the hive and collects everything that falls through the screened-bottom board (see "Hive Materials" section for hive set up) from the inner activity of the hive. After gaining some experience, this tool effectively allows the beekeeper to monitor mite levels, hygienic behavior, small hive beetles, colony/cluster size, wax production, brood production, mouse activity, robbing activity, pollen collection, and more. And this monitoring can be done regularly throughout the entire year, giving tons of helpful information to the beekeeper without having to open the boxes.

Through the winter months this tray may be your only visual assurance that your colony is alive and well. Through the warmer months it is the best non-invasive tool for taking note of hive hygiene and health.

Please reference page 141 for pictures and helpful observations with the debris tray.

On Wearing a Veil

A veil is to protect the beekeeper from getting stung. This protection may be very important for enabling the beekeeper to maintain a calm demeanor while working with the bees. Some people are concerned about the pain, others about swelling, and others about allergic reactions. Whatever the concern may be, there is nothing wrong with wearing a veil. It is important to wear a veil if it helps beekeepers maintain the calm directed demeanor that is necessary to do their work. This can be especially true for new beekeepers who have enough to worry about!

On the other hand, the ideal relationship with the bees is one based in the equality of mutual vulnerability. The bees use a special body language to communicate with their beekeeper. A common example involves one bee flying straight towards your nose, stopping right in front of you, and buzzing loudly as they hover back and forth for a few moments before returning to the hive. This is a primary warning for the beekeeper—the bees are "saying:" "Do what you need to do quickly and then close the hive." How the unveiled beekeeper responds to this first warning, or the second, or third, is personal preference. A veiled beekeeper has a more difficult task in maintaining a clear channel of communication. It might be assumed that the first bee flying towards the veil would have been to sting. This is possible, but not known. Sometimes the bees actually just come to check us out and say hello, without giving any sign of disturbed warning. Thus, we reach the central consideration concerning the veil–the pathways of communication are muffled. We greatly reduce the ability to listen to the language of the bees, telling us to "go ahead" or "go slowly" or "go away."

If we can gradually learn to approach the hive without protection, we will become all the more sensitive to the life of the beehive. On the right day, and with the right attitude towards the bees, our presence is welcomed and celebrated by them. On the wrong day, or with a distracted mind or heavy heart, the bees will let us know that we'd better look for a better time to work together. Simply put, with a veil we tend to dull our awareness to the needs of the bees. We are making a compromise in honor of our own needs. Sometimes this compromise is absolutely necessary. The important point is that this compromise is taken consciously so that the awareness of the bees' needs and the capacity to listen and be vulnerable can continue to grow in the beekeeper.

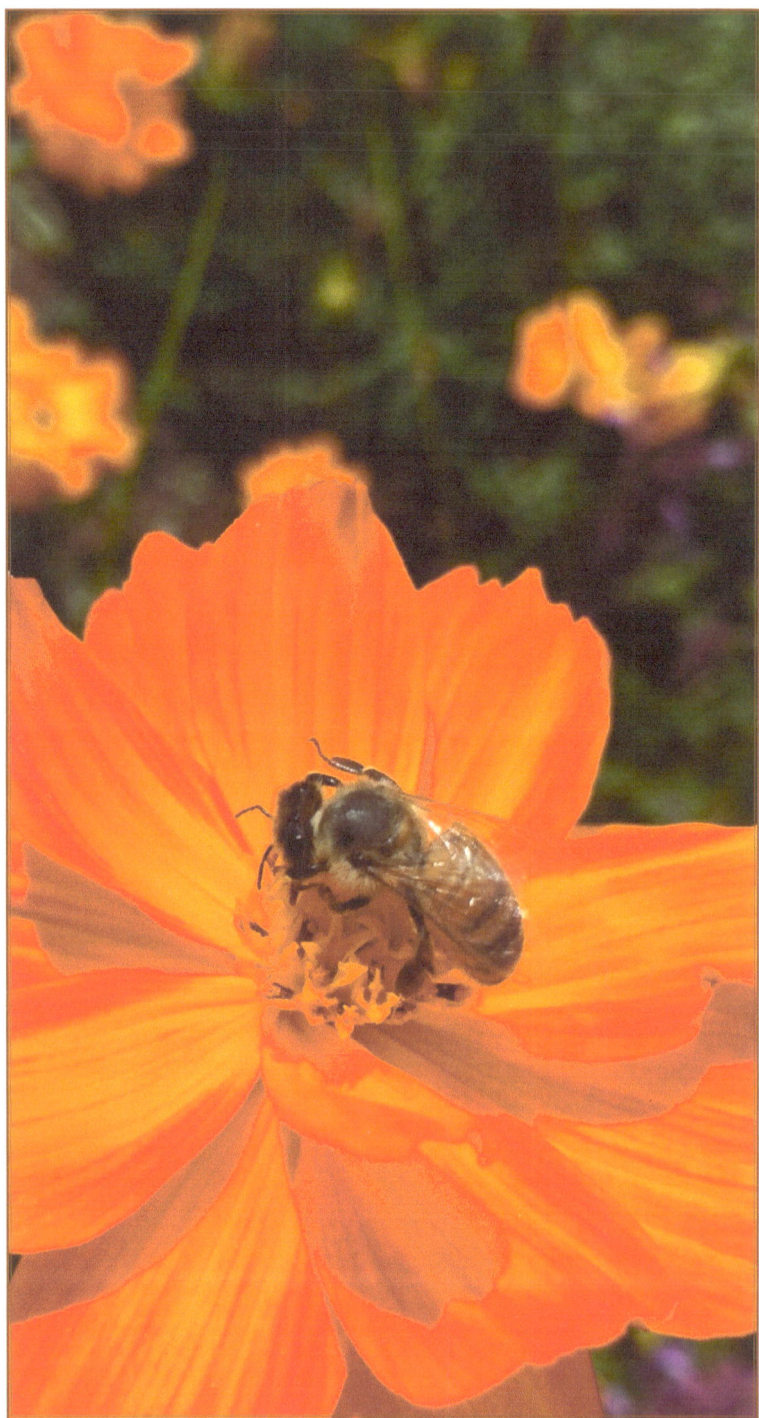

Chapter 2

How to Get Bees and Introduce Them

How to Catch a Swarm

The best source of bees, in terms of health, vitality, and local adaptability, is from a local swarm. Swarms are the best comb builders, have the strongest queens, and have a wonderful ability to make the best use of nectar flow and establish themselves as a viable and vibrant colony quickly. If they are from your own apiary or from close by, they already know where the good forage is available and can make use of it right away.

Catching a swarm can vary from an easy 30-minute process (the most common), to a multiple hour or even a multiple day process (quite rare). This is the part that can be intimidating, especially if you are getting into beekeeping alone and do not have someone close-by who is experienced in swarm catching.

Luckily, the principles and methods of swarm catching are quite simple.

If you don't have hives already that can produce a swarm for you, check in with a few beekeepers at your local bee club and see who might be willing to give you a call when they have a swarm for you to catch. They might even help you catch it if they have time.

Otherwise, a good place to build relationships in the community is with the exterminators and the fire department, both of which will often receive "swarm calls" from frightened citizens. These are often already related to a "swarm list" within your local bee club, so check with them first.

So let's say you get a call from someone who has a swarm of bees hanging six feet off the ground in an apple tree. An ideal scenario for a newbie.

As soon as you learn of the swarm, you must go there immediately. If you can verify that the swarm JUST landed, you'll have the best chance at catching them. Swarms don't always "hang around." The agreeable and docile nature of a swarm that has just left the hive degenerates noticeably if they have been sitting for half a day or have been traveling from one place to another, possibly even for multiple days. The fact that they have not yet found a suitable nesting site is compounded by the fact that they begin to run out of food stores from their honey stomachs, causing them to become increasingly protective.

On the other hand, most swarms are incredibly pleasurable to work with, and stings are very rare. When they leave the hive, full of honey, they have a lackadaisical way about them, swaying in the light of the sun, swinging round and round in the sky and then forming a large cluster from which they begin to look for a new nesting site. Here's where the beekeeper comes in and gives them the best option available—a well-managed hive in a friendly environment!

The process for catching a swarm is as follows:
1. Shake the bees into the box.
2. Put the lid on the box quickly.
3. Turn the box over gently and place it lid-side down on the ground below the swarm site.
4. Prop one corner of the box open with a stick.
5. Shade the box from the sun with a plywood board.
6. Wait. You might have to cut the branch and shake it at the "entrance" of your swarm catcher box after step 5 if there is still a sizeable amount of bees going back to the swarm site.
7. If most of the cluster is in the box in 15-30 minutes, brush as many bees as you can into the box and close them up. If the bees have returned to the swarm site, repeat steps 1-6.
8. Store the collected swarm in a cool, dark, quiet place overnight and through the following morning.

When a swarm leaves a hive, it looks like a pot of milk that was left to boil over on the stove: the bees come pouring out of the hive entrance and take flight.

Swarms typically gather into a beautiful cluster within 50' of their mother hive.

Swarm catcher box. We make these boxes out of 3/8'' baltic birch plywood so they are nice and light and easy to hold by the handle. They are 40 L volume, and the same dimensions as a Langstroth deep box (19 7/8'' x 16'' x 9 5/8'').

The swarming bees are usually very docile and easy to handle. Sometimes we gather handfuls of bees to help them collect in the swarm catcher box.

Especially on sunny days, we recommend using a plywood board to shade the swarm catcher box, which will make it more comfortable for the bees to gather inside.

Setting Up a New Hive

Ideally, all of your equipment (See Chapter 1) is ready and all you have to do is assemble your hive in the location you've picked out.

Introducing a Swarm Into Their New Home

Keeping the swarm overnight in a cool, dark, and quiet place has a wonderful unifying effect and allows the bees to strongly collect into a new individual entity. It improves the calmness of their demeanor and their sense of well-being, and the success rate of introduction into a new hive is greatly increased.

It also gives the beekeeper plenty of time for preparation. The new hive and stand must be made ready (see page 7) and the honey-tea mixed (see page 44).

Ideally, the beekeeper would be able to wait until the following day in the late afternoon to introduce them into their new home. This is weather dependent—you don't want to put them in while it is raining or during any other difficult weather. If it's going to rain in the afternoon, put them in before the weather system comes through.

Before you bring the swarm to your hive set-up, you'll want to take off the inner and outer covers and remove the five middle frames so that more of the bees will be able to fall right into the cavity. (Note for top bar hives—you will have to remove all of the bars because there is no "bee space" between them, as there is with the Langstroth frames.)

Reduce the entrance to the hive to about half so that

the bees don't come pouring out, but leave enough room to allow the bees who will fall down the sides of the box or fly into the air to have a chance to be guided into the hive by their sisters who will be fanning at the entrance.

The Knock-In Method:

With introducing a swarm, you often have to think on your feet a little bit, but the goal is quite simple: get as many bees as possible into the hive on the first shake and use a brush to slowly guide the rest of the bees in. Bring your swarm to the hive, and slowly lift the box from the lid. Most of the bees should be hanging together in an upper corner of the box or sometimes right in the middle. It's good to look under to see where most of the bees are sitting so that you position the box most advantageously when knocking them in.

Then, take courage, and give one swift and deliberate "knock," by bringing the four sides of the bottom of the swarm box in contact with the four sides of the hive body. One good hard knock will jar the bees from the swarm box, and they will fall down into, or onto, their new hive. Use a brush to guide bees from the side of the hive into the cavity. Once most of the bees are inside, slowly and gently glide the inner cover from the back edge of the hive body to its resting position, allowing the bees remaining on the edges to move out of the way. Then quickly put on the outer cover (or first your jar of feed with the feeder box, see feeding description page 44) before the bees begin to land on the inner cover. This will seal the top and the bees will have to begin using their front entrance. You should see bees from the inside coming to the entrance, and soon there should be bees fanning at the entrance to call the remaining bees inside.

While you do not have to see the queen, it is necessary that she goes into the hive body, or else the rest of the bees will not rest content inside and will fly out again. Many bees fanning at the entrance is an excellent sign that you have the queen inside the box and the bees are going to stay.

The Pour-In Method:

All the procedures followed above for the Knock-In Method will be followed here, as well. The only difference is that instead of knocking the bees into the box by making contact with the other box, we will in this method be knocking the bees into one corner of the swarm catcher box and then pouring them into the open space we have cleared by removing the empty frames. To do this, we gently flip the swarm catcher over, remove the lid, and bring the swarm catcher down to the ground. By thrusting one corner of the swarm catcher box to the ground and making firm contact, we jar the cluster loose from holding onto the box, and they gently tumble into the corner of the swarm catcher. Sometimes we need a few knocks against the ground to get the majority of the bees into the corner and to let go of their grip on the swarm catcher box. Once they are jarred loose and in the corner, we lift the swarm catcher to our hive body and pour the bees into the open space we have created (by removing a few frames in the center of the hive).

The Walk-In Method:

If you have a little extra time, warm dry weather, and maybe some friends that you want to share a magical moment with, the walk-in method provides a very nice option for introducing a swarm into their new home. This method is preferred at Spikenard over the knock-in or pour-in methods, especially for Sun Hives, Top Bar Hives, or any other hives where the equipment is advantageous to have in place beforehand, rather than trying to piece it together while the bees are trying to settle in.

To let a swarm walk in, you set up your hive and equipment, with your feeder jar or honeycomb already in place and everything ready to go. Put on your entrance reducer and open it up to the widest setting. Use a plywood board (or something like it) to make a ramp, which can be slightly inclined up towards the entrance or could be level to the entrance. Knock the bees onto the ramp. Have a feather or a beekeeping brush handy and watch closely. We are looking for the queen within the cluster of bees. If we find her, we may only need to watch to ensure she is going in the right direction—that she moves towards the entrance and into the hive. If this is not the case, or if she is just hanging out without moving much, you can use the feather to help guide her towards the entrance. Once the queen has entered the hive body, workers follow soon behind and will begin fanning her scent and their homing pheromone in order to give the rest of the bees directional instructions. Soon after the queen enters the hive, the orientation of the whole cluster shifts towards the hive entrance, and they begin marching in. This is a wonder and a joy to observe! It usually takes about 30 minutes for the majority of the bees to enter. You can always help them towards the entrance with your feather or brush if you want to speed up the process.

INSTALLING A PACKAGE INTO THEIR NEW HOME

Many people who are starting out for the first time will choose to begin their beekeeping journey with packages. I would recommend starting with two or three. The vitality of packages is not always very high, so the more resources you have in your apiary the better. Comparing and contrasting developmental observations is also a wonderful help and a good learning tool.

Packages are convenient because they are shipped through the mail, usually in bulk through your local beekeeping club. They should arrive in April in a temperate climate, or May at the latest. The package consists of: about three pounds of bees; a queen, who will be in a separate cage with a couple of workers who can feed her; and a can of sugar syrup.

As these packages are made from bees in many different hives, the queens and the bees originally have no relationship to each other. This does begin to happen while the bees are in transit, but in general, they have much less of a central, organized, individuality than a swarm does. For this reason, the introduction of a package is slightly different than the introduction of a swarm. With a swarm, you treat the bees and the queen as one unit. With a package, they are still sep-

Three pound package of bees.

arate, with the queen in a cage and the rest of the bees still coming together as a unity, both with themselves and with the queen.

When you get your package home, plan to install it into your prepared hive as soon as possible. With a swarm, we wait until the late afternoon to introduce the bees. However, the stress of their situation in a package, with the syrup and caged walls, is something that you'll want to relieve them of quickly. Have your hive set up beforehand so that you can introduce your package soon after you arrive home.

To install the package, first pry open the wooden cover on the top. Keep the wooden cover intact and with you because you'll need it soon. After the cover is removed, the syrup can and a tab that is attached to the queen cage will be exposed. With a pair of pliers, grab the upper lip of the can and pull it up and out. Shake the bees off of it and set it aside. Then pull the queen cage out, also shaking the bees off of it. Cover the exposed hole again with the wooden cover while you examine the queen and check that she is alive in the cage. After determining that the queen is alive, remove the cork plug on the candy end of the queen cage and with a thin nail, poke a hole through the center of the candy. Be careful—do not stab the queen! Then place the cage into your pocket.

Now you are ready to shake the bees inside the package into their new home. The bees will be crawling inside all over the package and all over each other. The first thing to do is to jar them loose from the walls so that you can pour the bees out of the upper hole where the syrup was once sitting. Hold the package firmly on either side and deliberately knock the bottom of the package against the ground. Then quickly remove the wooden cover from the top and flip the package upside down so that the bees pour into the empty cavity of your hive. You will have to shake the package back and forth a little until you get most of the bees out and into the hive cavity. Once 75% of the bees are out, repeat the process of knocking them on the ground and pouring them in to get as many out as you can. Then set the package down right next to the entrance of your hive. Take the queen cage out of your pocket and set it gently down, face up, on top of the bees that are on the bottom of the hive cavity. Then set your frames back in place slowly without pressing them down into place. As the bees move out of the way the frames will glide into place. Once the frames are in, take your brush and move as many bees from the outside of the box and the outside edges down into the cavity as you can. Then take your inner cover and slide it on slowly and gently. Take your jar of honey-bee-tea and place it over the hole of the inner cover. If the lid does not completely seal the hole, place a towel or handkerchief over the hole so that it is completely closed and the bees won't explore the space above their hive cavity. Take an empty deep box and place it above the inner cover in order to form a cavity where the feed jar will fit. Then close up with an outer cover once you've brushed any remaining bees out of the feeding cavity.

All of the remaining bees must now find their way inside the colony through the front entrance, which should be reduced to about half. Within an hour or so, all the bees will have found their way inside and maybe will come back out again to begin their orientation flights.

INSTALLING A NUC INTO THEIR NEW HOME

A nucleus colony, or "nuc" box, is another option for getting started with bees or expanding your apiary. Nucs are produced by beekeepers by various methods, and you want to be careful to check into the production methods before purchasing, especially because nucs are often twice as expensive as a package or a swarm.

Most nucs are fit for a Langstroth deep hive body and contain five frames that are managed for the first month or two by the beekeeper before they are sold. The small colony that is growing in the nuc box should have expanded their nest to five frames with brood, and of course the frames should also contain some pollen and honey stores. But the most important thing that you want to make sure you get is a nuc that has five frames with brood. This means the colony is of sufficient size and you are getting the full worth of what you are paying for this new colony.

The second thing to check for is how the beekeeper manages queens for the nuc boxes. If the beekeeper is making a split, this should be done only at the appropriate time, when the colony that is being split is raising true queens. You want to ask the beekeeper if it is possible to purchase a nuc with a true queen, rather than a grafted queen. Many beekeepers make a split whenever they feel like it and introduce a grafted queen into their nuc.

The other option, which is preferred but often rare to find, is to purchase a nuc that started out as a swarm. You might be able to ask a local beekeeper to do this for you by asking early in the year, in the "embarrassing" case that the bees swarm, if they would please make you a nuc box from a swarm.

The rarest thing to find in a purchased nuc is someone who is allowing the bees to build their own natural comb. Often, you will have plastic foundation that you have to deal with, which will take about two years to phase out of your colony. (See page 52 on working with natural comb.)

What is important to note is that you do not lecture the beekeeper from whom you are purchasing the nuc about biodynamic beekeeping and try to explain all the things that are wrong about conventional beekeeping practices. Do the best you can to find a good source, but do not be afraid to rescue bees from an otherwise unhealthy situation. Then gradually do all that you can to ennoble their lives and give them the care that they need.

When transferring the nuc to your equipment, the procedure is quite simple. You want to keep the core of the brood nest together for warmth and cohesiveness, so the center three frames in the nuc box will become the center three frames in the middle of your deep box. Then place an empty frame without foundation on either side of these three frames, and follow that with the remaining two nuc frames, keeping the left one on the left and the right one on the right. You now have seven frames in. Fill the remaining three spaces with three empty frames. Because you've set up a nice "checkerboard" pattern, the bees will build their new natural comb straight down the frame into the empty space.

Nuc to deep transfer:

FEEDING THE BEES

When introducing a package or a swarm to a new hive, the first major task for the bees is to build their wax comb. The bees make their own wax comb by metabolizing honey or nectar and transforming it through their fat body into wax. It takes seven pounds of honey to make one pound of wax! But in turn, one pound of wax comb can hold 23 pounds of honey. This wax comb is where the queen lays her eggs and the brood is raised, making it the essential foundation for the growth of the colony.

To support the growth of a colony in the early stages of development, we provide a swarm or package with supplemental honey and/or Bee Tea right away (as mentioned above, right when the bees are introduced to their new hive body).

If you have a comb of honey, one of the easiest ways to get a hive started is to simply provide them a nice heavy honeycomb in their new cavity, and they can just take the honey out of the cells and use it as a stimulus for wax building.

If you don't have any honeycomb, then a liquid feed is recommended. Honey is the best food for bees. Their own honey, or from hives in your apiary, would be the first choice. To list the bees' honey preferences in order:

1. Honey from their own hive.
2. Honey from the same apiary.
3. Local, bioregional honey derived from a similar landscape of nectar, organic or biodynamic if possible. Often best to purchase at a farmers' market or through a

direct relationship to the beekeeper so that you can consider their beekeeping practices.

4. Organic or biodynamic honey from a different region.

You will need at least ½ gallon, and possibly up to a gallon of honey per hive that you are feeding, depending on the nectar flow at the time of introduction. In this method of feeding, we dilute the honey a little bit to make it more liquid and also to imbue the honey with a good tea that can stimulate the metabolism and promote healthy digestion in our colonies. This tea we will henceforth call Bee Tea. This is a mixture of healing herbs that we grow biodynamically and sell at the Honeybee Sanctuary, and it is also something that you can grow yourself! The Bee Tea recipe includes these dried herbs:

7 parts of stinging nettle

9½ parts chamomile

2 parts lemon balm

2 parts dandelion

2 parts yarrow

1 part thyme

1 part sage

2 parts peppermint

When we make a tea with these herbs it infuses whatever we are feeding the bees with helpful medicine for their metabolism. Bee Tea and honey are mixed together in a proportion of seven parts honey to one part Bee Tea, which makes an excellent stimulus for comb building that is also nourishing and supportive. We add one tea bag full (about one heaping Tbsp) of Bee Tea per quart of water and let steep for 10 minutes. Let the tea cool down until it is cool enough to drink before you add it to the honey.

Access to such quantities of really good honey can be difficult and costly when you are just getting started. If this is the case, then a small amount of honey still will be necessary, but we can bring the quantity down to one pint of good honey per hive during the feeding period by using pure white organic cane sugar. We still need the honey so that the bees can more easily digest the sugar, which, if fed alone, can have detrimental effects on the bees' gut health by lowering the pH of the digestive tract and weakening the immune system, making the bees more susceptible to diarrhea and dysentery (nosema). With added honey, we are enriching the sugar solution with the right bacteria, yeasts, and ferments that are necessary for the bees to convert the feed into a decent food without getting sick in the process. The salt in this recipe also helps to make the sugar more digestible.

Here's how to make the Bee Tea recipe with organic white sugar:

1. Make the Bee Tea as instructed above.
2. Fill a small mouth Mason jar with the desired quantity of sugar.
3. Make a mark on the container (with a marker or a piece of tape) at the level of the sugar and add the hot tea, hydrating the sugar to exactly the same level.
4. Stir vigorously.

5. When cool enough to drink, stir in 10% of good honey.

6. Add a pinch of non-iodized salt.

Rather than using an entrance feeder or a tray feeder, we recommend using a Mason jar with a feeding lid (small holes poked into the lid) and feeding your hive internally. This prevents the attention of other bees and insects that comes with entrance feeding.

Refill as often as the colony finishes the feed for about two weeks—longer in rainy or cold weather, shorter during a big nectar flow. You know you can stop feeding if during your 10-day health check (The First Health Check page 74) you see capped honey in the cells. This means they have enough and no longer need the feed. If the feed sits on the hive for days and is not taken by the bees, then either they don't need it or it has turned sour (this can happen in warm weather). Check your feeder jar every two days until you establish a rhythm that matches the pace of their consumption and can keep them feeding consistently.

Feeder jar lid.

Feeder box in Langstroth hive.

Feeder in Top Bar hive.

Feeder in Warre hive.

Chapter 3

The First Three Weeks

GREETING THE BEES – A DAILY TASK

To actively engage in greeting the bees with interest each day can be one of the most rewarding and helpful practices as a beekeeper. My habit is to walk to each hive every morning. I observe the entrance, note the hum, smell their aroma, check that everything is in place, say hello, and breathe a few breaths while listening.

This check-in is the basis for the personal relationship that I have with each hive, and it becomes the doorway through which a deeper awareness develops. After doing this every day and comparing one hive to another, you get a really good sense for the well being of a colony. Predicting swarms, noticing the development of drones, seeing signs of a mouse, observing that they need a little more space, finding a dead queen—these are just some of the important life processes that a beekeeper can be part of by observing the hives on a regular basis.

51

Other benefits of "making the daily round" include:
- Giving you time each day to think about the bees and what they might need.
- Allowing the bees to get used to people coming to see them regularly.
- Getting the bees acclimated to the beekeeper's scent and presence.
- Observing skunk presence in the apiary.

Working With Natural Comb

The wax comb is an essential "organ" within the colony, and it has an incredibly diverse functionality. It is the bones that fill out the empty cavity of the hive body and give the colony its organizational structure. It is the womb where the queen lays her eggs and the nurse bees raise the young larvae. It is the fermentation chamber where nectar is transformed into honey and pollen into bee bread. It is the pantry where these foods are organized, stored, and protected. It is the dance floor where the bees do their waggle dance to communicate forage locations to their fellow workers.

The bees make this comb themselves and arrange the cell size, cell location, comb width and height, etc. all in relation to the cavity that they are provided and in relation to the needs of the hive and the availability of forage. If a new colony, say a package or a swarm, is to grow at all, they must first focus their work on the build-up of wax in order that any of these processes can begin to take place.

For the bees to create wax comb it is necessary that they

consume honey. The honey is transformed in the body of the worker into wax. She sweats out this wax from her abdomen, grabs the platelet with her mid-leg wax spike, and passes it up to her mandibles that are then used to manipulate and mold the pliable wax to its purpose. This metabolic process of the bees transforming their honey into "fats," which they then sweat out, is an incredibly revitalizing and healing activity. Especially for package bees, or bees from any other conventional source, the healing potential of wax creation is great.

In order to aid this healing process, Spikenard has developed our Bee Tea, which is suggested for use any time you are feeding the bees. If you ever have to feed sugar to the colonies because you don't have enough honey to spare them, the tea is a great digestive stimulus that helps the bees overcome the lack of nourishment in the sugar. (See section on feeding sugar page 42.)

All packages and swarms should be fed upon introduction into their new hive body for at least one week, often longer. They can take this nourishment and use it directly to build new wax. Check the colony that you are feeding after one week to see if they have any capped honey yet. Capped honey is the signal to the beekeeper that no more feeding is necessary.

At the Honeybee Sanctuary, we never use foundation because we have the resources from years of natural wax creation to help new hives get started on the right foot. It must be acknowledged that using wax foundation might be the easiest and most effective method for ensuring that your bees begin building their comb straight on their wood-enware. Aside from the state regulations, which often re-quire all hives to have removable frames, it is very import-ant for the beekeepers' relationship to the hive to be able to lift frames in order to see what is going on. But getting the bees to build straight does not mean that we need to use full sheets of foundation. Foundation should be minimally used at most. I will not go into further details of that here due to Hauk's important presentation on wax and foundation in *Towards Saving the Honeybee*. Suffice it to say that any kind of foundation is a crutch, at best, for a growing hive. At worst, the conventional foundation is a chemical time bomb and one of the main causes of hive toxicity and colony collapse. Even home-made foundation has no benefit to the bees themselves, although it has the potential to allow for a more fluid and pleasant working-together between the bees and the beekeeper—that is to say, it is purely for the beekeeper's needs that foundation makes any sense in this relationship. More on this later.

So, options for working without foundation must be ex-plored in bee-centered beekeeping. There are many ways to go about working without foundation, but only the methods that we use at the Honeybee Sanctuary will be outlined be-low. Our experience has led to a simplification of many of the methods that are used by other natural beekeepers.

After each season, we evaluate the wax combs that have been removed from our hives, saving the golden and white wax combs that were used for honey storage as full frames of comb that can be filled with honey by the beekeeper in order to feed the bees, or can be filled with honey by the bees when they are expanding in the spring and summer. The dark combs that have been used by the hives for brood rearing are treated differently. These combs are not reused except for the top inch that is left attached to the woodenware to help give the bees a guide for straight building. Once you have your hives building straight, you can continue to give them combs with a small strip of old wax in this fashion and they will follow suit. This strategy generally takes two years to establish, wherein the beekeeper must patiently cull combs and build resources from various hives in the fall contraction time (see Chapter 6 on Contraction, page 119) or post honey harvest. But combs can only be reused in this fashion if they are built straight by the bees in the first place!

Preparing the Equipment For Working with Foundationless Frames:

A nuc box transfer into a deep hive body, when "checker-boarded" as described in the previous chapter, makes it very easy to guide the bees to build their combs straight. Little, if any, additional help is needed.

For a swarm or package, it can be more difficult to encourage straight building without a little help when you are starting them off in a box with 10 empty frames. To make a compromise and use foundation at this point is acceptable. All that is needed is to buy a package of medium brood foundation, which will be the first and last package of foundation you ever need to buy, and cut the sheets of foundation into one-inch strips.

Wax foundation.

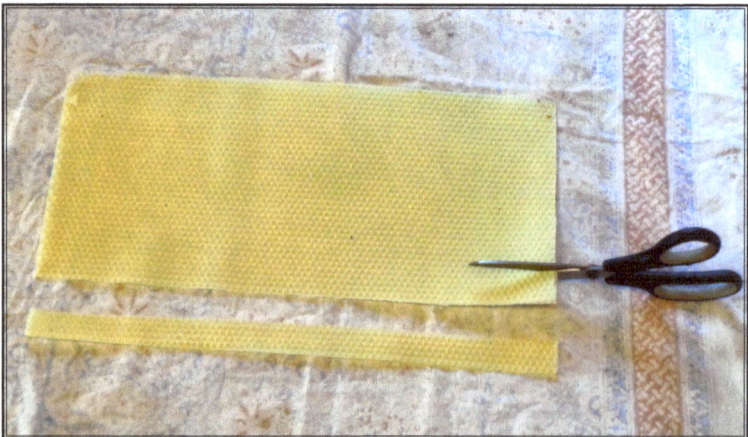

Cutting wax foundation into strips.

Attaching wax to frames.

The ideal would be to make your own foundation or to get together with your natural beekeeping group and make a plan so that good foundation can be available for beginners to get started.

There are options for improving your woodenware that can also help to encourage the bees to build straight without the use of foundation. Top bar hives, for example, are foundationless from the start. They utilize the knowledge that the bees will start building from the top down and hang from the lowest point to begin building their comb. Therefore, the top bars are constructed with a v-shape cleat, or ridge, that runs down the center of each bar, which encourages the bees to build straight down the cleat. This can be dipped in beeswax to entice the bees to go there and begin working (sometimes they chew off the wax, sometimes not, but at least it guides them to the point where you want them to begin drawing comb!). The same model has now been adapted for Langstroth and Warre hives, where the frames, or bars, can be purchased or made with a cleat that can help guide the bees to build straight along the frames from the lowest point. Before this invention, we used to take popsicle sticks and insert them vertically in the frame channel for the same effect.

These procedures only need to be employed until there is a sufficient supply of extra frames or bars with comb in your apiary. Once the hives have been established, there will always be extra combs that can be prepared to help guide the bees to build straight year after year. At the end of each season, when we are tightening and contracting the hives for winter, any superfluous frames that are not being used by the bees are removed. These frames are sorted—some are re-used and some are cut out to melt down. On the top of each frame, where the comb is going to be cut out for melting, a one-inch strip of wax is left as the guide for straight building.

Leave a one-inch strip on the top of your frames as you cut out old dark combs to give back to the bees as a template for straight comb-building.

No matter how you decide to set things up for yourself, some basic techniques in straightening out cross-combing is an absolutely necessary skill that you may have to employ— maybe as soon as your third day of being a beekeeper! In working with natural comb, especially if you make the decision not to use any foundation whatsoever, it is necessary to be prepared for comb-correction with each visit to your hive through the spring and summer. This necessity is most important in the early development of a colony and becomes less and less of an issue once the bees have established a pattern of building their combs nice and straight.

Day Three Hive Check

After the bees have been introduced into their hive, three days later you should go in to check on their comb building.

(If you have installed a package, this is also the day to go in and make sure the queen has been released from her cage. If not, check that she is still alive and carefully remove the screen and make sure she crawls out of the cage. Take the cage away from the hive area because it smells like the queen, and you don't want to confuse the bees. If the queen died, call your supplier immediately!)

The comb at this point is white and malleable, and will not be so large that it is impossible to manipulate. It becomes more difficult to correct cross-combing with larger combs. But by day three, the comb has generally not been drawn deeper than four inches down. This is perfect for correcting.

If you have given the bees a guide to build from (foundation strip, old comb strip, popsicle stick, cleat, etc.) then you probably won't have to do any correction. But you still need to check! If you find the bees have built their combs from one frame to the next frame, it needs to be corrected as soon as possible because all the other combs on each side will follow suit, and you'll have a full box that you cannot work with. The goal of correction is to have each comb lined up straight on only one frame. If the bees have been building their comb from one frame to the next for three days, the comb will span two frames, three at the most.

To correct and straighten the comb building, use a sharp knife to free the comb from its attachment to all of the frames except one. Once it is cut free, the comb can be gently guided and bent over and in line with the frame you want it on. Then, with your thumb and forefinger, press the free hanging portion of the comb firmly and gently against the bottom of the frame to secure it. This will crush a few newly built wax cells, but the bees will rebuild it quickly.

In this first check on day three, you are mainly looking to make sure that the bees have settled into their new hive, that they are building comb, and that the comb is straight. There is no need to worry about seeing eggs, larvae, pupae, honey, pollen, etc. at this point. We will come back to the hive and check on the queen's activity in about one week: on the 10th day after the hive's introduction for packages, and on the 14th day for swarms (this will be explained in the section, The First Health Check page 74).

PREPARING TO OPEN A HIVE

One of the most important rules in beekeeping is to know, or at least have an idea, of what you are going to do before you do it. Preparedness is all! Having the right tools and equipment at your side while the hive is open is essential—you do not want to have to go running around looking for a knife during the day three check because you forgot to bring it with you and the bees have cross-combed. Imagine yourself doing these procedures before you open up the hive and bring any equipment you think you might need before opening the hive. If you are prepared and calm, it is much more likely that the bees will be prepared and calm.

Each time the hive is opened, all kinds of external influences flood into the bees' sensory system that were not present when the hive was closed. Light, wind, different temperatures, smells, noise, and whatever the beekeeper subjects the hive to. Your thoughts and feelings are of consequence—the bees see right through you and experience your soul life. Calm thoughts, a warm heart, and directed actions can make for a very positive experience for the bees. There are many hives where we experience a great joy and contentment from the hives when we open them. They are happy to see us! But this relationship must be built and established with each hive. If they experience a frantic, nervous, or fearful beekeeper who is jerky in his or her motions and is indecisive, and if they experience this every time the hive is opened, they will begin to develop an aversion to the relationship. Opening the hive could become increasingly difficult, and defensive bees will only exacerbate the negative disposition of the beekeeper. On the other hand, if you start from the very beginning with a courageous attitude, commit to developing a positive and loving relationship with the bees through inner and outer preparation, and have a deep respect and reverence, then both the beekeeper and the bees will benefit greatly and a relationship can be established that can be nothing short of life-changing. The bees will speak once the beekeeper learns to listen.

Developing our Sense Perception

The bees are always giving us lots of physical information, and it takes time to develop a sensitivity towards the health of a hive in relation to how it sounds, smells, looks, or how warm it feels. What we perceive by observing the bees' activity can also help us determine the pervading mood of the day, which is often related to external factors such as the weather, the surrounding activities, and the planetary relationships. On top of that, each hive has its own personality and temperament related to both nature and nurture. The more of these observations that can be made before the hive is opened, the better. I find there is no better way to attune myself to the bees than to focus on my senses beforehand—observing the sky, smelling, feeling, and breathing the air, and watching the entrance of the hive intently for a few minutes.

It should also be noted here that the differences in hive shapes play a significant role in what we are able to see and how often a hive needs to be opened. A Sun Hive, Warre, or round hive of any sort tend to call for less interaction inside the beehive on a regular basis—often just twice a year. Even though our capacity to look with our eyes into the physical structure of these hives can leave us feeling unsure about the hive's general well-being, it can be an incredibly gratifying practice to learn how to identify health based on the sound of the hum, the smell of the hive, or the consistent observation of activities going on at the entrance to the colony. The practice of learning to trust our senses can be a wonderful training for the beekeeper's perception and intuition.

Below are a few examples of common sense experiences. This is certainly not a comprehensive list and therefore leaves you in need of continuing your own personal research.

Sound

By knocking on the side of the hive while your ear is near the hive entrance, or even pressed to the wood, you will easily be able to tell if the bees are there or not. By listening carefully, you can also tell if they have a queen or not. A healthy hive with a queen, after a swift knock, will buzz together, their hum rising and falling sharply in unison. If you don't have a queen, the buzzing will not be so united. The loud buzz at the beginning will draw on more slowly, and then the buzzing will linger, and then dwindle, then loud, then quiet, then a few more, etc. A queen-less hive SOUNDS different.

When you crack the outer cover on a hive with the intention to go inside, listen to the bees' reaction to their lid being opened. Often it is a unified, pleasant hum in unison that ends quickly, as if you gave them a knock (just mentioned above). But sometimes, it is a roaring and sharp reaction, maybe even accompanied by a bee flying out and buzzing around your face before going back into the hive. This re-action should give you pause from your intended work with the hive—the bees are telling the beekeeper that today is not a good day. In that situation, the best thing to do is to say "thank you," close the hive, and plan to come back soon. The question is not only about getting stung, but it is really about what is best for the bees. Even if you don't know why they are reacting in that particular way, it is best to trust and respect their unwillingness to be laid vulnerable on that par-ticular day at that particular time. Working against their will has the possibility of causing lasting damages to the trust be-tween you and your hive. However, there will be cases where you need to forgive yourself and go ahead anyway with the heightened consciousness that, for whatever reason, today is not the best day. This consciousness is already a big help.

With a well-trained ear, you don't actually have to knock on your colony to tell if they are there in the wintertime. Listening at the entrance, a low hush can be heard from a well-clustered colony, "shhhh...." Compare this almost inau-dible hush to a much louder hum in the wintertime and we may learn that a colony has an entrance that is blocked by dead bees, or it is too small for the hive to get enough fresh air inside. The bees can die of suffocation if they don't get enough fresh air in the hive as they exhale carbon dioxide. By cleaning out the dead bees or opening the entrance a little further, you should notice that by the next day the sound of the hive has turned into a calm hush.

Smell

A healthy hive has the most delicious and enchanting aroma—like a freshly baked sourdough loaf of bread made from the finest flours. Warm, rich, and heavenly. This smell is a great part of what makes up the hive scent, which is of central importance to the bees' well being. It is helpful to know the smell of a healthy, vibrant hive and use this as a point of reference for when we experience other smells coming from a hive. Each new nectar source gives a different aroma when the bees are fanning and ripening the nectar into honey. Particularly noteworthy is the goldenrod nectar, which can fill the whole apiary with an overpowering, warm, musty, slightly sour aroma that can sting the nostrils.

Smelling a hive is easy to do, especially if you have set your hive up as was described in the previous chapter, with a screen bottom board over a solid bottom board. You just open up the back as if you were going to take the debris tray out, stick your nose in, and take a whiff. I have done this before and perceived an aroma that was tinged with wet, mold, and coldness. This was a sign to me that the hive needed to be opened as soon as possible and tightened up! (More about this is explained in detail on page 119 Contraction chapter.)

Another important smell to recognize is the warning pheromone, which smells the same (at least to me!) from one hive to the next. When you accidentally surprise the bees either at the entrance or by opening the lid abruptly or without smoke or opening the hive on a day when they don't want to be disturbed, the bees will lift up their abdomens to the sky, extend their stingers, and send out a warning signal that alerts the hive of a potential threat. This smell is fruity and acidic, often comparable to the smell of pineapple or ripe bananas.

Sight

Utilizing or developing a keen eye for detail is an indispensable tool for beekeeping. As we'll see in the next section on record keeping, our noteworthy observations are most dominantly informed by what we see. The most basic physical observations when opening the hive include seeing eggs, larvae, pupae, brood pattern, wax building, etc. The activity of the workers can be watched carefully to gauge their calmness and general well being. Regularly viewing the activity at the entrance of a colony can help get a sense of the vibrancy and strength of a colony, especially when compared with what the entrance activity of another colony looks like. It is common for a hive that is queen-less, for example, to have workers that seem to be much more agitated and fidgety in their movements than normal.

As explanations of what we see in the beehive span every chapter of this book, I will not go further with those observations here but move to the less common practice of viewing the activity at the entrance of a colony.

Environmental Conditions on the "Right Day" and the "Wrong Day"

The bees are "in their element" when the weather is warm, sunny, and dry. In an emergency situation, we'll open up a hive when the temperature is as low as 55 degrees F, but this is to be avoided if at all possible. Overcast days are not necessarily unfavorable, but a change in air pressure due to oncoming precipitation will often make the bees uneasy. Certainly, the hives should never be open in the rain unless they are under a roof and it is an emergency. By 10:00 a.m. the majority of the forager bees have begun their work for the day and the hive is quite preoccupied with all of their daily activities. From 10:00 a.m. to 2:00 p.m. is the best window of time where the bees seem the most at ease with the beekeepers and more focused on their tasks than concerned about what we are doing.

Planetary Aspects

One of the unique aspects of research at Spikenard Farm and within the field of biodynamic agriculture is the attention given to the activities in our solar system and the stars. The relationships and influences of the sun, moon, and planets as they make their rhythmic cycles through space have increasingly been shown to have a close and intimate relationship to what is happening on our Earth. Working with these cosmic rhythms is both an ancient art and a modern science. For thousands of years, teachers have shed light on the relationships between the honeybees and the sun, the heavenly source of warmth and light, as well as the planet Venus, the goddess of love. Simple phenomenon, such as the fact that the bees pour back into the hive during a solar eclipse, or that the sun takes exactly as much time to rotate once around its axis as it takes a drone bee to develop (24 days), or that the bees seem to be consistently agitated when Venus is in retrograde show basic correlations that we can measure or directly perceive.

Working with and attuning ourselves to astronomical events and information is an activity that we take into account along with local weather conditions, environmental disturbances, our own inner disposition, and other conditions that affect our beekeeping practice and relationship with the bees. While a further discussion is not within the scope of this current book, I mention it here to open a door for anyone who might be interested to learn more. My first recommendation is to get a copy of the *Stella Natura Biodynamic Planting Calendar*, the *Maria Thun Biodynamic Calendar*, or any other localized biodynamic calendar, where astronomical charts are available for each day of the year and can help develop a relationship to this interesting field of study and observation. At Spikenard Farm, we are most notably tracking the activities of the sun, Veaxisnus, and the moon and their relationships to the life of the honeybees.

Hive Personality and Temperament

We name each of our hives, giving them the first letter of the mother hive that they came from. For example, if Fae swarms, we will name the new daughter hive starting with the letter "F," such as Felicity. In this way, we easily keep track of the lineage, and we develop lines of honeybee stock, each with their own signature, differences, and idiosyncrasies. Each hive has their own unique personality and temperament, their own way of relating to the beekeeper, their own history—whether they are sensitive, untrusting, aggressive, calm, friendly, accommodating. Also such aspects as the size of their brood nest and honey stores, the way they build their wax comb, or the yearly pattern of development and many more traits, differentiate one hive from another. These give the beekeeper the opportunity to learn the many different aspects of the various honeybee families and, through time, develop a loving partnership that is supportive to the special needs of each hive.

RECORD KEEPING

A pen and an organized notebook are just as essential for the beekeeper to use with skill as a hive tool. Notes give us a firm basis for approaching a hive with the knowledge of what was seen and experienced the last time the hive was opened. This can greatly help us to approach our work with the bees with all the possible tools and equipment we might need so that we do not have to scramble for a new box with frames, for example, while the hive is opened. Keeping records can also help us to get a picture of a hive through a full season and, ideally, through a long life and relationship together. It is exceedingly helpful and interesting to have notes from one single hive going back through seven or 10 years. Where did her lineage come from? How has she changed over the years? What has remained the same? What have we learned about her relationship to collecting honey, building her nest, dealing with pests, etc.? How long has it been since she swarmed? How old is her queen? All of this and much more can be answered and explored with the help of detailed records.

The biggest help in having good detailed notes is when the hive runs into problems and the beekeeper has to troubleshoot. (Please see Chapter 9: Troubleshooting Problems page 151.) Having as much information as possible is necessary for good problem solving because with the bees there are so many possibilities that must be accounted for!

Which brings us to the noteworthy fact that record keeping, when done faithfully and with enthusiasm towards trying to understand the life of a colony, can help us to learn and grow in our beekeeping practices and attitudes in a way that no book can teach us. Our conscious participation and astute observation of all that we take in through our beekeeping experiences is invaluable. There is no better teacher for a beekeeper than the bees! The goal, from my perspective, is to remove all barriers, veils, and crutches between beekeeper and beehive and allow for a positive and conscious relationship to unfold and be experienced.

Using a Smoker

The bees don't like surprise visits. Strive to give as much warning beforehand as possible. Tell the bees (really, go outside and tell them!) when you plan to open the hive days before you do it. The whole experience will be much more pleasant for both you and your bees if you imagine in detail what you are going to do before you do it.

Then, right before you enter, use a smoker to "ring the doorbell," as Hauk always says. Two or three puffs of smoke

Commonly Made Notes and Observations:

Objective	Observation	Note written down
Checking for brood health	Eggs, Larva, Pupa	Brood in all stages
Checking brood pattern	Queen seems to be laying eggs in the middle of the comb and working towards the outside. Brood in all stages is seen, and it is organized in a circular and regular fashion from frame to frame	Beautiful brood nest!
Checking strong hive in late spring for expansion	Lifting upper super—feels very heavy. Lifting lower super—feels light, looked from below without pulling frames. Upper deep—tons of bees, brood nest. Lower deep—brood nest	Upper super—heavy and full of honey Lower super—light, looked from below and saw 2 frames of honey with bees just starting to build new comb on 6-7 frames. Upper deep—full of bees, pulled one frame, saw brood in all stages. Lower deep—looked from above, did not open, tons of bees, seems vibrant! Action: Check on wax building and if she needs new super in 1-2 weeks.
Adding a new box	She has two deeps and one super and is ¾ through building the new super.	Added new super above deeps with two-frame bridge. Now she is two deeps, two supers
Adding a new box?	She has fully drawn comb on 5 frames, partially drawn on 2 frames, and 3 empty	5 full frames of comb, working on 2 frames. Still has room to build.

in the front entrance gives the bees a heads-up that you are going to open the hive. It is a training for them—each time the beekeeper is going to open the hive, the bees receive a few puffs of smoke at the entrance. Even new beehives get used to this very quickly.

What is used in the smoker fuel is quite important. Dried aromatic herbs, cut into one inch pieces, are the best smoker fuel—lavender, sage, thyme, holy basil, Echinacea, catnip, mint, hops, etc. We take the woody growth from perennials that we cut back, dry it, and chop it up for smoker fuel. It smells like a heavenly incense, and the bees seem to enjoy it too. It makes a nice cool smoke, which is important. If you have a hot blazing fire made of pine needles or wood chips in your smoker, and you are puffing embers into your hive, the bees will become nervous rather than calm. Also, burlap is commonly used, but is not well suited for the bees. Burlap is often laced with chemicals as harsh as rat poison for the shipment of goods, and the toxins that are given off when it burns are not something that I would recommend subjecting your beehive to.

How to Light a Smoker

Start with an empty smoker. Have a few 1" x 6" pieces of black and white newspaper (or something else non-toxic, like butcher paper), matches, and your smoker fuel ready at hand. Light the paper from the bottom and place it into the empty smoker. While gently puffing the bellows with one hand, begin sprinkling in handfuls of smoker fuel. Continue sprinkling the fuel into the chamber until you reach the top

of the smoker, puffing the bellows all the while. You should have a nice smoke erupting from each puff of the bellows, but the smoker fuel that you added should have put out any fire already. Press the whole mass of smoker fuel down forcefully into the bottom of the chamber, and keep the bellows pumping. Pump intermittently to keep lit.

If the smoker is needed for 10 hives or more, then you press down and pack in the smoker fuel tightly, all the way to the top of the chamber. I would recommend that you bring a little extra smoker fuel with you when you go out to the bees, just in case you need to add more. After a few visits to your bees, it will become easier to gauge how much you need to pack into the chamber to make it through all of your hives. Then you can begin to adjust, depending on how many hives you plan to visit on a given day.

Aside from giving a puff of smoke at the entrance, we will also use the smoke once the hive has been opened, sometimes just giving a puff to help clear the bees away from the woodenware so that the frames can be more easily handled, or sometimes if the bees get very agitated for whatever reason and we need a little help calming them down as we put the hive back together and exit. The only consistent use of smoke is as the doorbell. Otherwise, it is based on feeling and used as needed.

The First Health Check

The first look into the hive to see if the queen is beginning to lay eggs is always exciting. In this first real health check we also want to look to see that the workers are storing nectar and pollen and that they are continuing to build their comb straight and quickly. The timing for when to do this first health check depends on if the hive began as a swarm, package, or nuc.

Swarm

Primary Swarm: The primary swarm is the first swarm of the season for a given hive. These are often the biggest swarms, and the swarming bees leave the hive with the "old" queen. These queens typically have a very strong bond with the workers, and the swarms tend to be very well organized and get to work quickly and diligently when put into their new hive body. Roughly half of the bees leave the hive with the primary swarm.

Secondary Swarm: A secondary swarm is the second swarm of the season for a given hive. These are often smaller than the primary swarm, and the swarming bees leave the hive with a newly born queen, who is not yet fertilized. Roughly half of the remaining bees (who stayed behind during the primary swarm) will leave with the secondary swarm. Secondary swarms are slower to get established because the bond between the queen and workers is relatively new, and the queen has to go on a nuptial flight to be fertilized before she can lay fertile eggs, which happens after the swarm is introduced into their new hive body. If the nuptial flight is successful, the newly fertilized queen will return to the hive with the sperm she needs to begin laying fertile eggs. Depending on the weather, it can take two weeks for the new queen to begin egg laying.

Tertiary Swarm: A tertiary swarm is the name for any swarm that comes after a given hive has already had their secondary swarm. Tertiary swarms are the smallest swarms, and have a new queen with the same qualities that were described for the secondary swarm.

It's ideal to know if the swarm was a primary swarm, which would be the first swarm of the season from a hive and therefore contain the "old," fertilized queen. If this is the case then there is the possibility that you could see eggs being laid already when you go into the hive for the three-day check. In any case, we go into these hives for the first real health check on the 10th day to check on the activity of the queen. Often, we will be able to see eggs and larvae by day 10 with a hive that came from a primary swarm.

With a hive that came from a secondary swarm, we have the delayed timing that comes with the need of the new queen to go on a marriage flight so that she can receive sperm from local drones. This flight is weather dependent and cannot happen in the cold or the rain, but in the heat of a nice sunny day. In the case of a new queen and a secondary swarm, we wait to go into the hive (after the three-day check) until the 14th day to give the queen a chance to go on her flight, for her ovaries to develop, and for her to begin laying eggs.

Package: A package that is bought with a fertilized queen can be treated like a primary swarm, with the first health check on the 10th day.

Nuc: Because nucs should come as an established beehive with a laying queen, brood in all stages, developed frames of wax, and some honey and pollen already, it is necessary to check all of that right away when the nuc is transferred into the deep box (see Chapter 3, Installing a nuc into their new home page 41). After they are in their new home, their first work will be to expand their nest and food storage by building new comb on the empty frames that have been provided. 14 days is long enough to wait before going in again to see how the bees have been building and working in their new space. I recommend being ready with another deep box with frames at that time, just in case they have been diligent and are ready to expand.

Chapter 4

Spring and Summer Expansion

Honeybees have one of the most remarkable capacities in the household of nature: they are capable of creating wax out of their own bodies which they then use to build their nest. In comparison to their other Hymenopteran relatives like the ants who tunnel into the earth to make their nest cavities and the wasps and hornets who gather wood to create their paper pulp cells, the honeybees metabolize honey and sweat out wax platelets from their abdomens, and then sculpt these platelets into hexagonal wax cells. This metabolic transformation of honey into wax comb is the basis for the growth and expansion of a honeybee colony.

These wax platelets have fallen onto the debris tray. One honeybee can sweat out approximately eight wax platelets per day, and it takes just under one million of these platelets to fill one deep frame with wax comb.

The height of these comb-building activities coincides with a strong flow of nectar in the bees' environment. This season of growth will be referred to as the "comb-building season." When flowers are abundantly blooming and their nectar is flowing, the bees work to expand their nest and grow their combs into the extra space in their hive body by sweating out wax and forming it into new cells, which can then be filled with brood, pollen, or nectar. This comb building is the only way for a hive to grow the size of its nest, provided that there is room available for more wax comb to be built within the hive body. Without wax, there is no place for the queen to lay her eggs, no place for the workers to ferment pollen into bee bread or nectar into honey, and no place for the waggle dance and countless other communications. The wax comb is an essential organ in this superorganism, providing the womb, the fermentation chamber, the storage pantry, the dance floor, and the bones and structure of a colony.

Reference https://honeybeenet.gsfc.nasa.gov/Honeybees/Forage.htm for North American list of honeybee forage plants by month and their value as nectar/pollen sources

During the comb-building season, we want the bees to have just the right amount of room to continue growing. If there is too much space, their growth and capacity to keep a healthy nest structure can be negatively affected. Too little space and the beekeeper runs the risk of hindering a hive's growth potential or of forcing them to swarm. Of course, from the biodynamic perspective, it is not a problem if the colony swarms. However, a swarm that is cast due to lack of space was stimulated to do so for a different reason than the hive that has enough room to continue to grow, but was inwardly mature and strong enough to reproduce. The latter swarm is the result of sexual maturity; the former is due to a constraint of space. But again, too much space for the bees can be just as much a problem as too little space. In the extra space within the hive cavity that the bees do not permeate with their life and activity, we run the risk of having increased condensation, mold, varroa mites, hive beetles, wax moths, and other problems, too. Keeping the hive in a healthy relationship to the hive body space is important work for the beekeeper, especially when working with hives that are designed to allow the colony to expand and contract, i.e. Langstroth, Warre, Top Bar, etc. With hive bodies that have a fixed cavity space such as a Sun Hive or a tree trunk hive (30-60 Liters in volume), we prefer to introduce a large primary swarm (with an old queen, rather than a secondary swarm with a new queen) in the early part of the swarm season, that will be able to fill out the hive body within the first year so that they don't go into the winter months with extra unused space that would turn into a liability for them in the cold months.

Supplemental feeding (see page 82) of these fixed-cavity hives is sometimes necessary to help the bees fill the whole inner cavity as much as possible with wax comb, especially when the swarm is first introduced and also for when any rainy periods occur during the comb-building season. It is also important to note that the more natural hive types mentioned above also have much more insulation, sometimes four inches or more in thickness, and have round cavities, both of which help to nullify any potential issues with condensation, mold, temperature fluctuations, and some pest issues.

In the comb-building season, the general rule of thumb for expandable hives is that when the bees are 75% through building comb in their given space, they are ready for expansion. But what is true for the spring nectar flow will not hold in the summer dearth, the fall nectar flow, or the winter. It is during this special comb-building season where the comb is built that the hive's inner framework for the rest of the year takes shape. Each region will be slightly different due to climate and nectar flow, but the general pattern is that the bees use the spring nectar flow to build comb, the summer nectar to maintain their hive's size and strength and to add diversity to their medicine cabinet, and the fall nectar flow for winter honey storage.

These guidelines should help you to make the decision as to whether new space (i.e. new box, new bars) should be added to expand the hive body or not. Through the spring, the 75% rule mentioned above will hold as a good benchmark for deciding to expand. Generally we visit the growing/expanding hives every 2-3 weeks during the spring months to see if they are ready for more space. But there comes a point near the summer solstice when the bees dramatically slow their comb building. Adding new space at this tipping point is often hit or miss. It is common for temperate climate beekeepers to add a new box to the hives at the end of June or early July, only to take it off again in August once they realize that the bees aren't going to fill it up with comb. With experience, such actions can be avoided, which will only be to the benefit of the bees and save you time and energy. Established hives will tend to end their comb-building activities earlier than new swarms that are just getting established. We would highly recommend supporting swarms that come later in the swarm season, i.e. close to the summer solstice or after, with supplemental feeding so that they have help filling out their hive body with comb as quickly as possible. This will give the new swarms a much better chance at coming into a healthy balance between the size of the nest and the honey stores as the season progresses towards winter. More on how to work towards a balanced ratio of nest size to honey stores is discussed in Chapter 6.

After the comb-building period has passed, a beekeeper can add empty wax combs that have been saved from previous years into a hive so that they can fill up these combs with honey stores for the times of dearth. This allows for the summer and fall nectar flows to be utilized for winter honey storage and can help to increase honey resources for the whole apiary (if they are not needed for overwintering in the particular hive that stored the honey). Wax combs can be stored and saved for this purpose from previous years, but they must be protected from wax moths if the comb is to be preserved. (See Chapter 5 for more information on wax moths.) This is done by building a comb rack in a sunny place and placing the combs roughly 2" apart so that light can penetrate into and shine through the combs through the course of the year. Wax moths are most attracted to the dark brood combs or to combs that are stored in darkness, but they are repelled by light.

How do those combs get so dark? As honeybees go through their metamorphic stages of larva to pupa, the larva spins a cocoon of silk around itself inside the wax cell. These silk cocoons are left behind once the pupa hatches into an adult honeybee, and after a few generations of honeybees one cocoon has been left on top of another and the combs become darker and darker.

By choosing and saving the nice yellow combs and spacing them 2" apart, we will almost completely eliminate the danger of wax moths moving in and decomposing the wax.

Honey frames must be stored inside in order to preserve the honey from being eaten by honeybees and other insects.

Empty wax combs can be stored outside, preferably on a comb rack in a south-facing shed where the light can penetrate between the combs, which prevents wax moths from moving in (but not wasps from building a nest!)

One of the benefits of having hives with expandable boxes and frames is that this allows the possibility for the brood nest to be built anew each year with fresh comb. Letting the bees build their own comb anew each year is one of those tasks that can help the bees in their adaptability and longevity. Each season, the nectar and pollen flows will differ as the weather, temperatures, and humidity differ, and the hives' inner needs will differ. By allowing the bees to build some new natural comb each season, we can serve their capacity to adjust the cell sizes of their comb to allow for more of whatever might be needed, whether it be more small cells for workers or pollen, or additional larger cells for drones or honey. In current and future times, with increased climatic changes bringing about less stable weather conditions, the need to continue to encourage and support the resiliency and adaptability of our colonies is essential.

The yearly exchange of combs in honeybee colonies has also become more important in recent history due to the high levels of pesticides and pollution that many beehives have to deal with in their environment. When pollen, nectar, or water is collected from a polluted environment and metabolized by the bees' bodies, many of the toxins are sweat out with the excretion of wax platelets, and these toxins accumulate in the wax comb. Therefore, in areas that are heavily burdened by pesticides, heavy metals, and other pollutants, we can keep the levels of toxicity within the colony at bay by exchanging the old dark combs in the brood chamber and allowing the bees to build fresh comb. The methods and timing for removing the older combs is covered in Chapter 7.

In hands-off, fixed-cavity hives, where the exchange of combs is not possible through the help of the beekeeper, the bees will naturally chew out and replace old combs over time, especially when the wax cells have gotten so small that the queen can no longer fit her abdomen in to lay an egg! The process of comb exchange in these hands-off hives goes slowly, which means that if your environment has high-toxicity and pollution problems, it might be more challenging to keep bees healthy in a hands-off, fixed-cavity hive. When the land is sick, the bees often need more care and intervention to stay healthy.

Because toxicity builds up in the wax, the recycled wax foundation that is so often recommended in conventional beekeeping practices is a poor choice if we are focused on keeping the honeybees healthy. Any kind of foundation, whether made from beeswax, paraffin wax, or plastic, is actually unhelpful for the bees in the long run. It is commonly stated that it saves the bees work and makes the hives more productive. The reality is that when you try to save the bees from working, their health with be compromised. The bees live to fulfill these activities, which we call work—it keeps them healthy! Hives that are in control of their cell sizes, are self-organized and able to freely sweat and metabolize, and are given the time they need to make their own comb are the ones that are most likely to survive and develop resilience and adaptability. It is commonly experienced that the hives that grow over-abundantly large in the spring and early summer are the ones that are often in the greatest danger of perishing. We call these hives "over-conditioned," a state that is so often encouraged because it produces hives that are outwardly huge and productive. Foundation is a crutch that produces the short-term gain of more honey production at the expense of the hives' capacity to survive. Maintaining balance is key for hive health, and anything we do to artificially stimulate the quick or early growth of the colony has potential for having adverse health effects later on.

From this observation we can modify the conventional goal of having bigger hives, and we can encourage well-organized hives in the apiary instead. So what does a well-organized hive look like?

Well-organized hives have an innate proportion and balance of their different parts: the workers, drones, queen, wax comb, and their hive body all make up essential organs of this superorganism. All are necessary and have their tasks which are essential for the functioning of the whole. We honor the bees' own instincts and wisdom when it comes to allowing them to build their own comb, raise their own queens, and reproduce through swarming. These are all essential to support for the healthiest functioning and growth of the whole organism from a wider perspective. To get down into the details, we can also see the reflection of how this wisdom and instincts manifest themselves when we look into the nest of a well-organized hive.

The healthy organization of the nest begins with the work of the queen. The activities of the queen are the center-point of the hive's inner life and give the whole colony its common direction towards the future through her egg laying and chemical signaling. In carefully observing the pattern of the queen's egg laying, as well as the development of the eggs into larvae and pupae, we can see how this brood nest becomes the center around which the rest of the hive is organized.

The queen begins laying her eggs in the center of the comb—not too low, high, left or right, but in the center! From this point, she works out towards the periphery, consecutively laying egg after egg in an expanding spiral. These eggs, and the larvae that hatch from them, have to be kept at a minimum of 95°F in order to incubate properly and metamorphose from egg to larva to pupa and into adult bees with healthy immune systems. Therefore, we want to support this capacity in our hives and also to make note of a queen that is laying her eggs in a spotty, disorganized, or otherwise inefficient manner. It is also excellent to note if the queen is actively utilizing the front and back of each comb and mov-

Eggs

Larvae

Worker pupae are under the capped cells in the left-half of the picture.

ing to adjacent combs to continue her pattern, not just in one small section, but allowing the whole nest to grow as the workers build more comb and bring in food stores.

From this warm center of brood rearing, the hive expands outward in layers of pollen/bee bread close to the brood, and then honey on the outer edges of the nest. Together, this forms an organization pattern commonly known as a "brood pattern."

A beautiful brood pattern should be celebrated and noted in the beekeeper's journal and is the first sign of a well-organized colony. A strong, healthy queen is of incredible benefit for the whole colony. Vibrant queens sometimes live for six years or more, leaving new hives each year in their wake. Swarms and the creation of drones from these queens can help to increase the productivity and health of the stock in the apiary and to improve the genetic stock of the whole landscape of honeybees.

We compare this beautiful brood pattern to disorganized and unhealthy patterns in Chapter 9 on Troubleshooting Problems, see "spotty brood nest" on page 158 for example.

The brood nest should be honored with great reverence. It is the most delicate and sensitive aspect of a colony, and its vulnerability to potentially harmful outside influences must be mitigated as much as possible by the beekeeper. The bees themselves often help to guide the beekeeper away from the brood nest if they would not like it to be viewed or manipulated in some way. It is much more likely to find defensiveness of this brood nest as the temperatures begin to cool in the fall or in a shortage of nectar—in both cases, when resource

A beautiful brood pattern. Note: capped drone pupae have raised caps on the left side of the picture, as compared to the flatter worker pupae cells.

conservation becomes the priority for the hive. Certain hives will be more protective than others. It is very important to learn this about the individual hives so that the proper approach can be measured and taken towards the hives in each season. For example, there are some hives where inspecting the brood nest in full can only be accomplished in the spring. The most that will be comfortably tolerated by some colonies in the late summer or fall may be the removal of one brood frame, just so that the beekeeper can be sure that all is well (that there are "brood in all stages:" eggs, larvae, and pupa).

Some hives call for us to learn to use and trust our faculty of intuition. We must learn to know that our hive is doing well without actually seeing the brood nest in some cases. How do the bees sound? How do they move? What does the hive smell like? Is there an abundance of bees? When was the last time we actually saw any sign of brood? In some cases we need to trust even if we can't physically see and then utilize our other senses to form a picture of hive health from a different set of information than is normally relied upon.

Learning Together at
Spikenard Honeybee Sanctuary

Chapter 5

Pest Control

The entire Spikenard Method of Biodynamic Beekeeping is built from the underlying foundation of respect for the bees' own needs. In serving her instincts, we are serving her health, balance, and homeostasis. We are often asked: what do you do for foulbrood? Or, how do you deal with nosema? Etc. Well, in over 40 years of practicing these methods, Hauk has never had one case of foulbrood or nosema. I can vouch for this—these common issues are simply not a problem at Spikenard Honeybee Sanctuary and for other beekeepers that practice biodynamic methods. But beekeepers who feed sugar, corn syrup, and pollen patties, who use artificial foundation, and who ventilate their hives excessively have many issues with viruses, diseases, and metabolic issues. Natural beekeeping eliminates many problems that are otherwise caused by the conventional beekeeping methods.

On the other hand, we must be very clear about the challenges that the bees are facing today. The honeybee is a domesticated animal that relies on our care. Today, more than ever, the environment presents many difficulties for a creature as sensitive as the honeybees are. Many of these issues can be helped by allowing the bees to live a natural life, but there will also be times when the 'natural' beekeeper will face a decision: Do I offer medicine to help the hive to regain homeostasis and heal, or do I hope that the hive will take care of the problem themselves? Our approach at the Sanctuary has created deeply personal relationships with our hives, wherein we would feel completely uncomfortable to take the hard Darwinist line and proclaim that any hive that is struggling ought to be allowed to go through their process without support.

So we do treat our hives with natural medicines on a case-by-case basis, monitoring the health and well being of the hive closely before the decision to treat is made.

Hive beetles, wax moths, and varroa mites are the main troubling insects we have experience with regarding the regulation of their proliferation. The pesky animals that we have encountered are mice, skunks, and bears.

Small Hive Beetle
Aethina Tumida

Small hive beetles lay their eggs in honeycomb. The eggs hatch and the larvae feed on the honey, leaving a trail of excrement behind them as they go which causes the honey to ferment and spoil. When the larvae mature, they crawl out of the hive entrance and go down into the soil next to the hive where they pupate and hatch into adults.

Hive beetles are most active in warmer climates, but as the climate changes their reach is creeping further and fur-

Small hive beetles on debris tray.

ther. When the adults enter the hive, their exoskeleton is so tough that the bees cannot use their sting to get rid of them. The worker bees instead pester and chase the hive beetles out of the hive, either out the front entrance, or out of the hole in the inner cover, etc. The bees will sometimes also use wax and propolis to make little "beetle jails" into which they corral the hive beetles and guard them, starving them inside.

In general, strong healthy hives will have no problem dealing with hive beetles. Keeping the colonies tight without too much extra space will eliminate places for the hive beetles to hide and will help the colony control them. Using vinegar or rock salt on the ground around the base of your hive will help to kill the pupae of any hive beetles that are in the ground and help to keep the population down. Little "beetle jails" can also be purchased which can help the bees to have more places to corral and starve the beetles.

When looking inside a colony, if you see an infestation of hive beetle larvae, those combs should be removed immediately and the hive should be tightened as much as possible. A large infestation is a symptom of bigger problems with the colony's health. When we open hives, we tend to squish any adult hive beetles that we see scurrying around. The last tip is that you can smell the fermenting honey by sniffing at the hive entrance—the sour, over-fermented smell can alert you to a problem inside and the need to help the hive as soon as possible. Because this fermenting honey becomes more liquefied, you will also see it drip down onto the debris tray and be able to taste/smell that it is fermented.

The last thing to mention is the good practice of straining your honey through a fine sieve during honey extraction to make sure you filter out any hive beetle eggs or larvae that might be in the honey. This is especially important if high hive beetle pressure is common in your apiary.

Wax Moth

Wax moths chew through wax comb. They especially prefer comb that has had pupae in it—nice dark combs where they can make their silky channels and lay their eggs. They don't really disturb honey cells. The wax moth is the natural companion of the honeybee as a recycler of unused wax. Their rightful and historic place inside the beehive is down at the bottom of the cavity where they eat and recycle the wax that falls down from the normal activity of a colony. Therefore, the few wax moths that are hanging around will be found on the debris tray or on the bottom board where wax platelets and gnawed wax fall down from the activity of the workers above. A strong honeybee colony that has

Wax moth larva on debris tray.

control over the whole of its hive body will have no trouble keeping wax moths on the periphery.

If wax moths are found to have infested the area where the nest should have been, or used to be, it is not because the moths forced their way into the hive and caused the hive to dwindle and weaken. Rather, it is hive weakness and the decrease in numbers of bees and the incapacity of the hive to extend their work into all of the hive body, leaving big areas where the comb is unattended that allow for the right conditions for the wax moths to come in and utilize the combs that the bees are not attending to for their work. In this way, the growing presence of wax moths could be an indicator that you may have bigger problems with your hive.

The natural contraction of hives, which is discussed in more detail in Chapter 6, will also naturally bring more wax moths into the hive during the season of contraction, the winter in temperate climates. But in the season of growth when the hives should otherwise be expanding, if the wax moths are a growing presence in the colony and in the combs, it is surely a sign that your hive needs special care. When you realize that the bees are not able to take care of their whole hive body and combs, the first thing to do is to tighten up the hive. Remove all the combs that the bees are unable to care for, leaving the brood nest small and tight and leaving them with no more honey or space than what they need. In this way, we can strengthen the hive simply by helping her to maintain greater control over her cavity.

VARROA MITE
VARROA DESTRUCTOR

The varroa mite has adapted for thousands and thousands of years with the honeybees of the Far East. It was only recently in the 1980s that the varroa mite made it to Europe and the 1990s when it arrived in North America. Its natural home is within the beehive, and unlike the hive beetle that reproduces in the soil, the varroa mite reproduces inside the capped pupal cells. So naturally, as the brood nest gets larger through the season, the varroa mite population will increase. And in late summer when the brood nest of the colony has reached its height and begins to shrink again, the varroa mite population will follow suit.

The varroa mite attaches herself to the thorax or abdomen on a honeybee and feeds on honeybee fat and blood (haemolymph). After feeding, the mite hops off its host honeybee, leaving an open wound. Honeybees with a strong immune system will be able to heal these wounds and carry on with their work. But in immune-compromised situations, where weakness in a colony has allowed the colony to be-

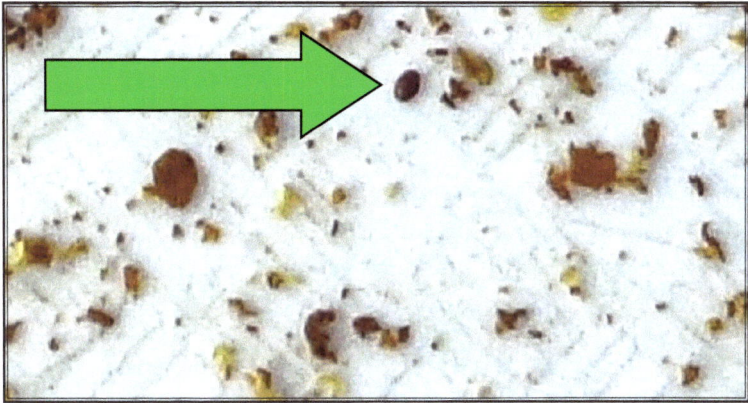

Varroa mite on debris tray.

come overwhelmed with a mite population that is growing out of control, we also face the risk of those open wounds becoming infected and these infections spreading. Often the secondary infections that spread through the colony are actually what end up killing a colony. Deformed wing virus, Israeli acute paralysis virus, and many other secondary infections commonly spread in hives with weaker immune systems following varroa mite infestation. Because the relationship of the varroa mite to the European honeybee is relatively new, and because conventional beekeeping practices have done so much damage to the vitality and strength of the honeybees, the losses of colonies that become infested with varroa mites is huge.

Monitoring the level of varroa mites with a debris tray is a weekly practice at the Sanctuary so that we have a clear picture of what the situation is inside the colony. As we mentioned before, it would be natural to count many more mites falling per day at the height of the season than when the nests are the smallest in the winter.

Varroa mite on a honeybee thorax.

Mite with dimpled exoskeleton.

Drone with deformed wings.

The mites that have fallen down on the tray over the course of the week give us a picture of the natural death rate of the mites per week in the colony. The number of mites that we count give us a data point which we consider as helpful information but only in relationship to all of the other observations, notes, life history, and general state of well-being in the given colony. 10 mites per day means relatively little to us out of context. For example, it is more concerning if we count 10 mites per day in February than in July. And it is more concerning if the hive has only one deep box rather than two deeps and two supers. And 10 mites per day is not very concerning if the numbers are staying level at 10 mites per day week after week. That shows you the mites aren't able to grow beyond the capacity of the bees' strength to keep them at bay. We often see mites that have been chewed in half, mites that have dimpled exoskeletons, and living mites crawling around on the debris tray. These are all potential signs of hygienic behavior in the hive—that the bees are coming into a relationship with the mites and are able to cast them out of the hive.

On the other hand, if the numbers are 10 per day this week, 20 per day next week, and 40 per day the next week, THEN we know that we have a problem that the bees need help to get under control. And if there is any sign of secondary infections, which can often be viewed by observing the dead bees out front of the hive entrance in the morning, we know that we have an acute problem that needs to be addressed ASAP. We will need to address the underlying weakness and strengthen the hive from a holistic perspective as well, but in this life-or-death situation we can save the colony by relieving the mite pressure with the use of formic acid.

FORMIC ACID TREATMENT

Formic acid is a substance that the bees "know" intimately. Formic acid is in their poison, their venom, and is the acid that is common to all of the order Hymenoptera, the stinging insects. Formic acid is naturally also part of the exhalation of the bees as they are breathing. Inside the beehive we already have formic acid permeating the atmosphere. So, what we are doing by applying formic acid within the colony is raising the normal level of formic acid beyond what the varroa mites can cope with but still keeping it at a level that does not adversely affect the honeybees. Extensive research has shown that this level is 65% formic acid. You can buy formic acid from a number of different sources, and it often comes in higher concentrations that need to be diluted to 65% with water. Remember that gloves and a facemask should be worn when dealing with this corrosive and flammable acid, and always add the acid to the water rather than the water to the acid!

Formic Acid Dilution (by volume)

85% to 65%	3 parts acid to 1 part water
90% to 65%	2.5 parts acid to 1 part water
95% to 65%	2 parts acid to 1 part water

We take 50 ml of this 65% formic acid and pour it over an organic cotton feminine pad. We open the hive lid and

Applying formic acid treatment.

place the formic acid pad between the frames and the inner cover on the fifth and sixth frames so that when we put the inner cover back on, the hole in the inner cover is sealed by the upper part of the pad. The formic acid then begins to evaporate and permeates the hive body. For larger hives with more than two deeps and one super, we place the feminine pad above the lowest super so that it is sitting above two deeps and one super. For Top Bar hives we pin the formic acid pad on the follower board or in between two bars.

In order that the formic acid does not evaporate too quickly or too slowly, the best temperature range (daily highs) for applying the formic acid is between 60°F and 85°F. If you are closer to 60°F, reduce the size of the entrance to a few inches wide. If the temperatures are closer to 85°F then open the entrance completely.

Keep the formic acid pad inside the hive for about one week before removing it. If you see a lot of the cotton down on the bottom tray after 3-4 days, it means that the bees are chewing the formic acid pad, which is a sign that the formic acid has evaporated and you can go in a little earlier to remove it.

If the treatment was successful, you should see a huge amount of mites that have fallen to the debris tray, sometimes 700 per day or more. The increased mite counts will continue for about two weeks. This is because the formic acid penetrates the pupal cappings, so when the adult bees emerge from their cells, the mites that died in there when the treatment was done will be cleaned out and will fall onto the debris tray.

Formic acid does not accumulate in the wood, in the wax, or in the honey. This makes it safe to use in relation to honeybee and human health, as well as makes it more difficult for the varroa mite to develop a resistance to it. Formic acid is also very effective for use against tracheal mites.

There also are commercially available formic acid treatments that you can buy from North American beekeeping suppliers, such as Mite Away Quick Strips.

BIODYNAMIC PEST ASHING OF INVERTEBRATES

The ashing of pests is an integrated pest management practice used in biodynamic agriculture. The method of pest ashing (sometimes referred to as "peppering") originates in the agriculture lectures given by Rudolf Steiner in June of 1924. Steiner presents his method as an effective process for working with planetary rhythms and spiritual influences to affect a positive relationship on the farm and in the garden by reducing pests that are proliferating beyond the manageable threshold. He gives indications for warm-blooded animals, weeds, and other pests, but the focus of this section is specific to two common invertebrates in honeybee colonies— the varroa mite (Varroa destructor) and the small hive beetle (Aethina tumida). Any invertebrate could be substituted in the following process.

In this method, pests are collected during peak fertility. For small hive beetles and varroa mites this is in the spring when strong honeybee colonies are growing their brood nest very quickly and the varroa mites and hive beetles are naturally following closely behind with their reproductive cycles. The pests can also be collected during other times of the year and dried so that they can be preserved until the ashing date in the spring. For ashing mites and small hive beetles you need about a teaspoon full of each, more is fine. (Avoid including bee parts in your collection.)

Steiner's research—which has been substantiated by several others, most notably Matthias Thun—indicates that we ash the insects when both the sun is in the sign of Taurus

(May/June) and the moon is in the sign of Taurus (for ~3 days every month). Taurus, the Bull, is the revered fertility symbol in many spiritual traditions and has a close association with the planet Venus. See the *Stella Natura Biodynamic Planting Calendar, The North American Maria Thun Biodynamic Almanac,* or similar calendars for date reference in relation to the astronomical configurations.

Just as water is the physical representative that brings life and fertility, fire is the opposing physical force that brings death and destruction. We bring the pests into a fire process under the sign of Taurus to help limit excessive proliferation. A fire is made in a metal container with wood that gives a good amount of solid embers. It is helpful to separate the embers into individual piles, one pile for each insect that you plan to ash. Enough wood should be burned to guarantee a handful of embers for each pile. Separate the embers into piles and pour the insects into the embers, gently poking them in between the coals. The ashes of the insect will be mixed with the wood ash.

Pest ashes.

After 10-15 minutes this ash is sifted from the solid coals and is put into a mortar and pestle and ground/dynamized for one hour. As the ash becomes stickier and gets finer, use a wooden spoon or popsicle stick to scrape it from where it gets stuck on the sides and continue grinding.

With this dynamization, the particle size has decreased to ~0.2 microns (the size of a virus) and has increased in surface area. A very little bit of this ash goes a long way. We apply the ashes right after we make them while the sun and moon are still in Taurus. We re-apply once/month when the moon comes back into Taurus. To apply, we take the ashes and put a little into each hive by smearing some on the debris tray and some across the entrance where the bees are entering and leaving. We also place a little in the soil near each hive. Gently pressing your fingertip into the finely ground ashes will give you enough to leave a nice smudge of ashes without using up too much. For some hive shapes, it can be helpful to roll your ashes in with a small ball of clay or Earth, and place it on the bottom board for the season.

Pest ashing.

As we dynamize the ashes, and again as we apply the ashes, we consciously carry the intention that the *pests do not increase beyond the capacity of the bees' strength.* By placing this intention into the work, we seek to bring the spirit into our process, into matter, and allow the intention to manifest for the good of all. We believe the pests have an important place on Earth and a home in the beehive, and we do not seek to banish them or destroy them but to help them to form a healthy homeostatic relationship between them and the honeybees. We bring our intentions to what spiritual and indigenous traditions have often called the "Group Soul" of the species— the Great Bee, the Great Mite, the Great Hive Beetle. We call on them to help us with our task of stewardship.

With these intentions, we make and apply the ashes which, after applied systematically for 4+ years, can greatly reduce the proliferation of mites in the apiary when aligned with the whole of biodynamic beekeeping practice.

MICE

Mice would love a warm, cozy beehive to build their nest in for the winter. When they do move into a beehive, it is not so much a problem of the mice eating all the honey as it is about the mouse feces which affect the whole hive's feeling of well being. The smell of mouse poop and pee is a great stress for the beehive, and the bees will orient themselves to sting and chase out the mice if at all possible. A large, robust hive will not have much trouble driving mice away, but the weaker or smaller colonies are sometimes unable to cope with mice.

The best thing we can do to prevent mice from moving into our colonies is to install an entrance reducer at the front entrance of the colony in the late summer before the mice begin looking for a place to make their winter nests. These entrance reducers need to be $\frac{1}{4}$ inch in height—enough to let the bees pass in and out but small enough to keep out adult mice.

The debris tray is a very easy place to check to know if mice are in your colony. You will see the nesting material, the poop, and gnawed bits of comb right away. If mice have moved in, don't waste any time! Look for the first possible moment when you can get in there to chase the mice away and clean up. It also might be a sign that your colony is weak and might have too much space. Putting on the entrance reducer or eliminating the entrance where the mice got in is the first step. Then tightening up the hive space as much as possible and feeding a good Bee Tea are both recommended. Then check often to make sure the mice have not moved back in.

Skunks

Skunks eat insects. They are very helpful with eating grubs in the lawn and keeping beetle populations in balance, so we honor their work there. But when it comes to beehives, skunks can do some serious damage. Skunks approach beehives at night and scratch at the entrance with their paws, luring out the guard bees. As the guard bees approach or try to sting the skunk's face, the skunk brushes the bees from its snout right into its mouth. Skunks are said to be capable of eating ¼ pound of bees per night, which, after several visits, can seriously diminish the population of a hive and cause lasting damage, not to mention the stress it adds to the bees! I have found hives that were incredibly aggressive during my morning visit, which accompanied observations that a skunk had been at the hive the night before. Scratches on the wood at the entrance and matted grass around the front of the hive are both indicators that skunks have been there. I have started putting little rock designs on the front of our hives, and when I see the rocks out of place or knocked onto the ground, I have another indicator that the bees have had a visitor. Skunks are more prolific in wet years when the beetle grubs rise closer to the surface of the soil, making an easier food source for the skunks. In prolific years when skunks begin to put pressure on the beehives, the only effective way I have found to solve the issue is by trapping and killing the skunks.

BEARS

Contrary to what we thought we knew from Winnie-the-Pooh, bears love to eat brood. They will open up a beehive and eat the proteinaceous brood nest with all the juicy larvae and pupae and leave the honey behind. Dogs can be helpful in keeping bears away—even just having their scent around your land is a benefit. A good 10-foot deer fence works for the lazy bears that would just as soon eat berries than climb the fence. But a determined bear can easily get over a deer fence. And they are often reported to go right through electric fences as well with their tough skin and thick coat. Baiting the electric fence with bacon or peanut butter can help, but it's not fail-safe. The two methods of bear protection that we recommend are building a bee shed or building a bear platform.

Bear platforms are very successful in evading bears. They are built 7 feet off the ground with the posts set in 3 feet under the edge of the deck. This makes it so that the bear cannot climb up to where the beehives are located above the deck.

This method of protecting bees is quite common, especially in Eastern Europe. The main features of these sheds include an easy flight path for the bees to travel in and out of the shed and access doors on the back side where the beekeeper can enter to work with the bees and check the debris tray.

Chapter 6

Contraction, Consolidation, and Looking Ahead to Winter

The bees follow the rhythm of the sun. In northern latitudes this rhythm and the relationship between the winter and summer solstice offer quite distinct polarities in the life of the beehive. As the sun is growing from the winter solstice towards the summer solstice, we have a time of expansion. And as the sun is waning from the summer solstice toward the winter solstice, we have a time of contraction. This should not be taken rigidly but more as a picture of how the hive is orienting itself. Each of these periods has a slow beginning, a steep rise towards a peak, and then a slow transition towards the opposite pole. The most distinct example of this is the hive's creation of wax comb and honey.

In temperate climates there is a distinct comb-building season at the height of the nectar flow of spring. At this peak expansion time the nest is undergoing tremendous growth and reaches a height that allows for the creation of drones and, later, the reproductive miracle of the swarm. The comb that is built at this time will later be fixed as the basis for the colony to organize their winter cluster and their honey stores through the lean and cold months. By the summer solstice comb building slows drastically, and the hives begin to transition into their new orientation towards winter.

This knowledge gives rise to a number of important considerations for the beekeeper. How do I want each individual hive to go into winter? When is the last time that I can add more space (new box, bars) and expect them to fill it with comb? How much honey does each hive need to go into winter?

It is of the utmost importance to leave no superfluous space in the hive—the ideal is that every frame or bar in the hive is fully drawn with wax comb and the wax comb is also full with nest activity and/or honey storage. Empty spaces and empty combs invite problems in the time of contraction. Mold, hive beetles, wax moths, and mice all love to move into the empty combs where the bees aren't extending their care.

Late summer is a perfect time to begin looking at the consolidation of hives and start removing empty frames and boxes. As the hives are checked at this time, consideration is given to both the size of the brood nest and the amount of honey that is stored. The relationship between these at the fall equinox should be 1:1. If you have a full deep box of brood, then you'll want to have a full deep box of honey above them to overwinter. If your lower deep has roughly 2/3 brood and 1/3 honey, then you can get away with having a full super of honey above the deep going into the winter. We are talking about a ratio because it is just as possible to overwinter a five-frame nuc by this method as a hive that has two deep boxes of brood to three full supers of honey. **Having too much honey can be just as much a problem as having too little.** If the bees do not need the honey, it will be extra space that they cannot warm, ventilate, and care for, and it will only cause problems with overwintering. It is always better to keep the bees tight at the 1:1 ratio and store the excess honey for feeding in the late winter/early spring.

Excellent note taking is essential for the bee work at this time. Where one hive is lacking, another may be able to share and help to equalize. Overwintering is the main reason why it is so important to have beehives of the same type in your apiary. The hives that have a standard size (Langstroth, Warre) are easier to share honey stores among than other hive types (Top Bar Hives, Sun Hives, etc.). By having at least two of the same type of hive, you will greatly increase the resources and possibilities of care for each hive. If one hive has too little honey, this is the time to share from a hive that has more than enough.

With this information, we can go to each hive and begin to evaluate their winter needs. We need to do this evaluation in the late summer—not too late—because we have the chance to actually do something to help each hive (or better said, to help them help themselves!) before the colder temperatures set in. Equalizing honey or sharing extra frames of comb for the bees to fill should all be done before the start of autumn so that stock is taken of the available resources in your apiary. Then you will have to make some decisions: Are all of the hives strong enough to make it through the winter? Do I have to feed a hive so that they will have enough honey? Do I have to combine a queen-less hive, or one that has a very weak brood nest, with a stronger one?

The goal of fall feeding is to allow your bees to take the feed into the hive, ripen it quickly, and add it to their winter stores of honey so that they arrive at the fall equinox with the 1:1 ratio of bees to honey.

In hives that have movable boxes and frames, tightening up is one of the beekeeping tasks that can greatly help hives make it through the winter. From the summer solstice towards the fall equinox we have an ideal time to begin thinking of how we can help organize the combs and boxes to support the dynamics of the winter cluster (see Chapter 7). Colonies naturally organize themselves with the brood nest closer to the hive entrance and the honey stored furthest away from the entrance.

For Langstroth hives this means that the brood nest will be in the lower boxes (deeps) and the honey is stored above them in the upper boxes (supers). The bees move UP into the warmth and into the honey stores as the temperatures cool and winter sets in. As the colony moves up, we are sometimes given the possibility of helping support the dynamics of the winter cluster by removing the bottom deep once the bees have moved up out of it. This is done on a warm enough day (above 50°F) anytime between November and March. We open up the hives on those days without pulling any frames but simply by tipping the boxes to look to see where the winter cluster is sitting and if they have moved out of the bottom deep. If they have moved out, we remove the box and brush out any remaining bees that will be passing through. Removing this superfluous space is a great help in supporting the colony to efficiently maintain their warmth for the coming months until comb building begins again and we can add the cleaned-up deep box back to the hive.

For Top Bar hives, reorganizing combs can be essential for winter survival, depending on the placement of the hive entrance. If the entrance is on the short side of the hive, then it is already well organized with the brood towards the front and the honey stores in the back bars of the colony. These colonies will move **BACK** into the honey stores as cold temperatures set in and winter progresses. As the colony moves back, we can pick a warm enough day, as stated above, and remove the combs towards the front of the hive that the bees are no longer clustering on or using, and then adjust all of the remaining combs forward to the front of the hive. If the entrance is on the long side of the hive, we might have to reorganize the combs if the entrance is centered in the middle of the sidewall. With centered long-side entrances on Top Bar hives, the bees will develop the brood nest in the middle of the cavity close to the entrance and will store the honey on both the left and the right of the brood nest. This creates a problem going into the winter, where the cluster has to choose to go **LEFT** or **RIGHT**. Either way they go, there is the danger that they eat themselves into a corner, and even though there is honey inside the hive cavity, the bees can still starve if the honey is all the way on the other side of the hive. Rearranging the combs so that all the brood is on one side of the hive and all the honey is on the other side can prevent this potential starvation scenario. In this way, the bees can move their cluster in one direction into the honey stores.

Hive Wrapping

The standard ¾" wooden boxes are pretty inadequate for overwintering a hive. Whether it's a Langstroth, Warre, Top Bar, or you name it, hive bodies are generally built way too thin! It is, of course, a matter of practicality that the beekeeper is able to actually lift the boxes. It is unfortunate for the bees that this challenge is not generally shouldered by the beekeeper, but many natural beekeepers are moving in a helpful direction. Hives with thick walls help the bees keep cool in the height of summer and warm in the cold of winter. Condensation and mold issues are minimized to a significant degree with more insulation, and the hives can much more efficiently utilize their honey stores through the winter. The Sun Hive, the Spikenard Hive, and tree trunk hives are all beneficial to the bees because of both their adequate insulation and the roundness of the hive cavities.

For those hives that still have 2" or less of insulation built into their hive body, we need to wrap the hives to provide extra insulation for the winter months. Hive wrapping begins when temperatures begin to drop below freezing at night. In Virginia the season seems to call for wrapping in early November. We do not wrap earlier because through September and October we still have warm days where we can manage honey stores and tighten up the hives as much as possible

before we wrap them. This is in consideration of conve-
nience and time efficiency so that we do not need to unwrap
the hives in order to check on them and then re-wrap them
again. We like to get the hives down to their ideal overwintering state—tight and full of honey in the 1:1 ratio—before
giving them a really nice insulation layer and allowing them
to pass through the coming months undisturbed.

The hives keep their wrapping through early April. It is
not so much the cold that we are trying to protect the bees
from but the fluctuations of temperature through the coldest
months. Steady temperatures, whether warm or cold, are
the easiest on the bees' winter cluster, and giving them good
insulation helps the bees maintain their homeostasis and buf-
fers them from the weather fluctuations.

1. First measure your twine — two pieces per hive is usually
sufficient.

2. Then measure your plastic sheeting, starting 3'' over the outer cover.

3. Place a brick on the outer cover to hold the plastic sheeting and roll the sheeting down to the hive stand.

4. Cut with your scissors at the bottom of the lowest box.

5. Open up the sheeting, holding one end in each hand.

6. Line up the middle of the sheeting to the front of the hive.

7. Wrap around the hive left and right

8. Use the brick to hold the sheeting in place and tie your twine around the sheeting, making sure the twine is above the hive entrance.

9. Remove the outer cover.

10. Place a handkerchief over the hole in the inner cover so that the bees and warmth stay inside the hive. Begin stuffing your hay or straw into the sleeve you have created.

11. Stuff hay generously on all sides—you can't really overdo it! 2-3'' is enough insulation.

12. Remover the handkerchief. Fold the plastic sheeting over the inner cover, and slide the outer cover into place, taking care that the plastic is tucked under the outer cover.

13. Tie your upper string around the hive to keep the wrapping more secure.

14. Place a few flakes of hay above the outer cover for extra insulation if you like. This can also be accomplished with a quilt box.

Horizontal hives are wrapped as well. We wrap them up like a burrito and then cut a hole at the entrance so the hive can breathe and fly on warm days.

Quilt Boxes

Quilt boxes are commonplace in Warre hives as insulation boxes that are placed above the hive where it is actually most important. We add quilt boxes on all of our hives and leave them on year-round. These boxes are the same length and width as the other boxes on their hive and are 2" tall or more. Organic cotton fabric or untreated burlap is stapled onto the bottom, forming a breathable box that is then filled with insulation material. In our trials pine needles worked the best in terms of insulation and mold-resistance. Other natural materials that work well include straw, hay, wood shavings, and wool. These materials should be checked once a month or so to make sure they are not wet or moldy, in which case they should be taken out and replaced with fresh material.

Quilt box on a Langstroth with Rye straw.

Quilt box on a Langstroth with pine needles.

Winter hives, all cozy and wrapped.

Chapter 7

Wintertime to Starving Time

The bees don't "sleep" or hibernate in the cold months but are in a very intimate time of inwardness. The way that the bees prepare and organize themselves in the winter cluster is a very important dynamic to understand. The task of the winter bees is to create and preserve an atmosphere of warmth within their tightly organized cluster that maintains a proper relationship to the hive entrance, to airflow, to humidity, to condensation and temperature within the hive body, and to the available honey and pollen stores in proximity to their cluster.

This warmth is especially necessary for the development of the brood. As long as the queen continues to lay eggs, the hive will need to maintain 95°F within the brood chamber regardless of the outside temperatures. If a steady cold winter ensues, the hive may take a break from raising brood. The queen can then rest from laying eggs, and the winter cluster can cool down to 65°F which is ideal for the conservation of energy and honey.

Each individual bee has the capacity to generate warmth by unhinging its wings from the flight muscles in the thorax and vibrating these muscles as if shivering. After about 30 minutes of generating warmth, the worker bee is exhausted and has to rest. She is relieved from her work by another who is subsequently relieved by another, and so on, until all of the bees in the cluster have rotated from the outside of the cluster to the inside and then back out again. The bees on the outside of the cluster form a sort of skin that helps to contain and preserve the heat that is generated by the group of shivering bees in the core of the cluster.

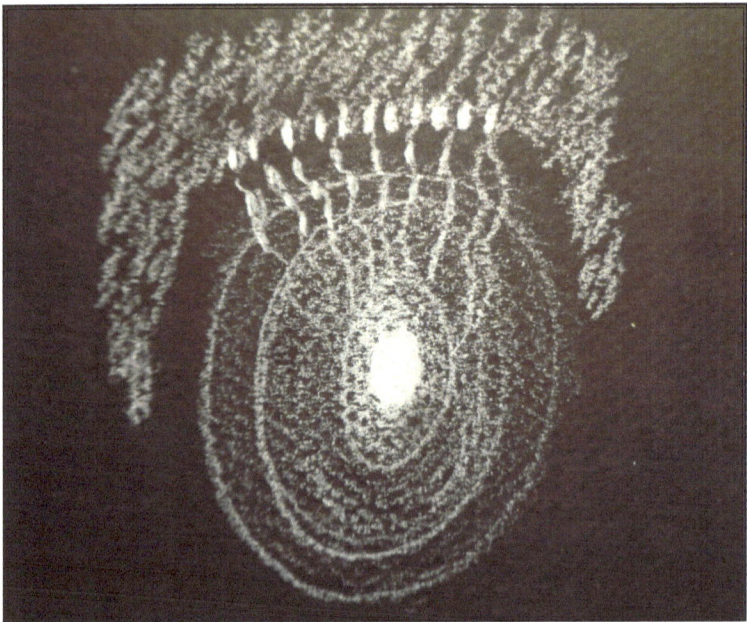

This diagram illustrates the warm cluster in the center where the queen is laying eggs, surrounded by layers of bees who are keeping in the warmth generated in the center of the cluster. Then we have the few bees at the top who are taking honey from the cells and then passing it from bee to bee to bee until everyone gets fed.

The colony consumes the honey that is close by and within the cluster first before slowly moving the cluster, comb-by-comb, towards the honey stores as the winter progresses. There are only a small handful of bees that actually go to the honey cells—their job is to take the honey in and then pass it off, like blood flowing through veins, from bee to bee to bee until all the bees in the cluster are fed, including the ones in the very center who are last in line but exerting the most energy.

The worker bee has the capacity to take honey in and then feed or pass the honey down the line. However, the drone bees do not have this capacity—anatomically, the drones' proboscis is suited only to receive honey, but it cannot give it up again. Therefore, a drone would be the clog in the artery of the flowing honey through the winter cluster, and all the bees in line behind the drones would be left unfed and hungry. Thus, the drones are let out of the hive in autumn to prevent their disruption of the winter harmony. Amazingly enough, the queen too, would be the weak link in the chain as she is also unable to give honey to another bee. Her position, however, is the last in line at the very center of the winter cluster—there is no bee that is waiting behind her to be fed!

The other side of the "bigger is better" attitude among beekeepers is the prejudice that a colony with less than 5,000 bees is too small to survive the winter. This may be true sometimes but not as a general rule. What we have found is that in order to survive, the bees need the correct ratio of 1:1, bees to honey, in a hive body that is as tight and well insulated as possible. A colony can create the same winter dynamic with a tiny cluster—we have seen a late swarm survive the winter with only a couple of hundred bees that we placed into a quart-sized, queen-rearing nuc and insulated it heavily. To see the tiny brood nest, come through the winter in early March—20 capped pupae, 30 larvae, 15 eggs—was such a joy! All they needed was ½ pound of honey to get them through to the dandelion season, and then they were able to grow into a strong and vibrant colony through the course of the year. The colony ended up living for two more years in a Top Bar hive.

Winter feeding should be done only in an emergency. There will always be hives where it is not 100% certain that they have enough honey to make it through until April, so those are hives that are candidates to check on a warm winter day. Any day that it is above 55°F and sunny provides good opportunities to check on the weaker hives to see if they might need continued consolidation or possibly honey. The very best method for increasing their food supply on these winter days is to have full frames of honey in storage that you can place down in the hive right next to the cluster. These honey frames can be stored in a safe place indoors or in a refrigerator. Some people will freeze the honey frames for fear of hive beetle eggs and larvae, but I would not recommend it. First off, the freezing doesn't always kill all of the eggs—they are hardy and often able to tolerate the cold. And secondly, much of the healthy, living aspect of the honey will be damaged through the freezing, and the honey will quickly crystallize, making it more difficult for the honeybees

Honeycomb storage rack.

to use in the cold months. The best option that has been successful for us so far is building a little hanging rack from the ceiling to store the honey frames. This rack should be in a room that has a lot of light, if possible, and a room that you can check on at least once per week to make sure no foul-play is occurring. This honeycomb storage rack is the same structure as the wax-comb rack that was outlined in Chapter 4, except that it is inside rather than outside.

If you do not have honey stored in frames to give the bees, but you have extracted honey, another good option is to use your finger to smear honey into empty wax comb, effectively making a honey frame that you can place into the hive right next to the cluster.

The last option is for the emergency times when it is just too cold to open up a hive completely and you know that they need food. In this case, you can make an emergency fondant paste to spread around the inner cover of the hive where the bees can go to feed. The recipe for the fondant paste is:

1 cup of honey

1 tsp strong bee tea

1 pinch of good salt

3 cups of powdered sugar

Mix the Bee Tea, honey, and salt together in a mixing bowl or a standing mixer. Slowly add the powdered sugar one cup at a time until you reach a nice stiff paste that does not run at room temperature. Place this into a piping bag and squeeze onto the inner cover near the hole.

Feeding fondant paste.

One of the most important monitoring tools for the wintertime in the Langstroth hive is the debris tray. We check the trays through the winter months about once every two weeks. Valuable observations can be made regarding the size of the winter cluster and the amount of activity within the cluster. Sometimes you'll see the bees eating themselves into a corner or the size of the cluster shrinking drastically from week to week. Both of these would give cause to find a nice day to open up and check to see if anything needs to be done to help them. It is usually late winter when you can begin to see wax platelets on the tray or sometimes even a few eggs will drop onto the tray—both signs that growth has begun internally.

1. As outlined in this book, our Langstroth hives are outfitted with a debris tray that can be accessed from behind the hive. Many different hive shapes can be outfitted with debris trays.

2. The little reddish-brown oval in the center of this picture is a varroa mite. Weekly monitoring and counting of varroa mites can help us stay attuned to their population curve.

3. It is not alarming to see a wax moth or two on your debris tray. This is their natural home in the hive—down on the bottom where they can recycle the wax.

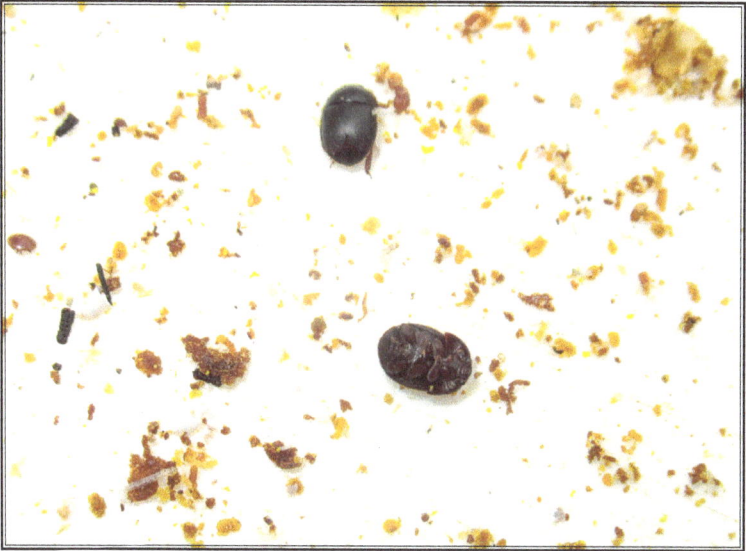

4. Hive beetles on debris tray.

5. These ants should be brushed off, and then baking soda applied around the edge of the debris tray and the back closure slot. Baking soda is helpful for keeping ants away.

6. This busy debris tray gives a picture of a hive that is big and strong and is permeating the whole hive cavity with its activity. Very nice!

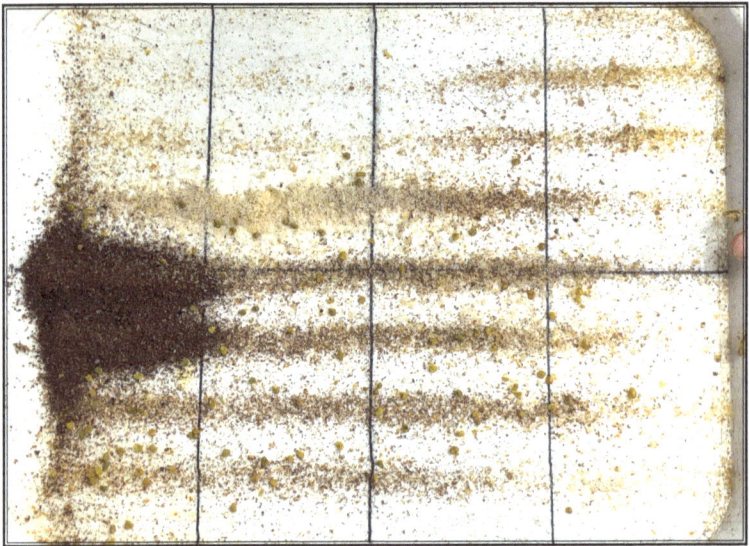

7. This winter tray shows a fairly large cluster who is chewing out a lot of old dark wax at the front of the hive. Sometimes this is because the wax is moldy. Sometimes the cells are too small or are the wrong form and the bees chew them out to renew them. This is often seen in the late winter/early spring.

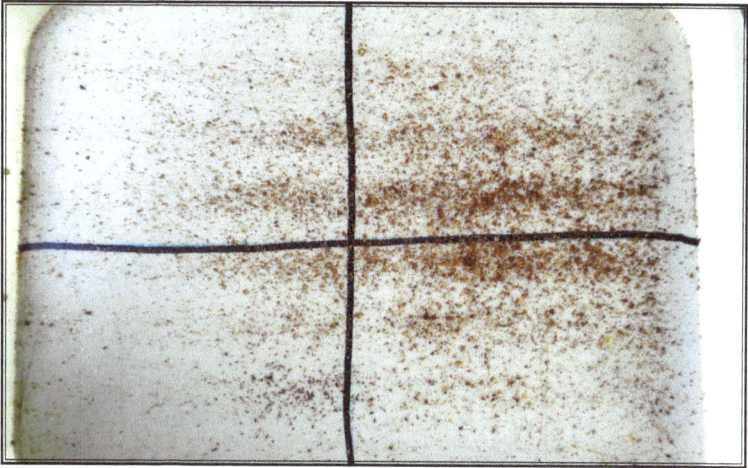

8. This winter debris tray shows a nicely sized cluster. If you clean the tray after every observation, you can watch how the cluster moves and changes shape. Usually, the cluster size gets smaller and smaller as the winter progresses.

9. This winter debris tray shows a very small looking cluster with concentrated activity. Its position at the front center of the hive means that it can most likely travel up or in any direction to find honey stores. If such a small cluster started moving into a corner over the course of the winter, it would be cause for concern. We would then look for the next moment to go in and check to see if there was enough honey, or if the hive needed to be tightened up.

10. Crystallized honey is often seen in the late winter, when the bees start tapping into honey stores that have been sitting far from the cluster all winter. The honey in the farther reaches can begin to crystallize, which the bees mix with water or nectar to try to liquify and use again. Some of this falls to the debris tray in the process.

11. When the queen begins to lay eggs again, around early February here in Virginia, one can sometimes find a few loose eggs that drop onto the debris tray.

12. When pollen is being brought into the hive and is unloaded, some of the little packages inevitably fall to the bottom of the hive and end up on the debris tray. This is a wonderful way to see the different colors and the diversity of pollen sources that the bees are gathering from.

13. When the bees begin to expand again and comb-building season begins with the dandelion bloom, one can see lots of wax platelets that fall onto the debris tray, which inevitably drop as the bees are passing them around and working with them to build new comb.

When the growth of the nest begins as the sun grows stronger in late winter, we are faced with a tender and potentially dangerous moment in the season of beekeeping. This is the starving time when the temperatures of the colonies need to be maintained at 95°F for the developing brood and the bees begin to consume a lot more honey in order to maintain the proper nest warmth. This inner growth is usually accompanied by outer changes in the season as we approach the spring equinox. As winter and spring dance back and forth together, the outside temperatures fluctuate back and forth. This is a big transition for the bees where they find themselves needing to maintain their warmth through these big fluctuations with potentially limited provisions. Some days the bees can fly and gather the fresh spring pollen and water to feed the larvae, only to ball up again in their tight cluster to brace against the cold of the nighttime or the following colder day.

The winter is an ideal time to connect deeply with the bees and cultivate a meditative relationship with each colony. While you cannot go into the hive and do your normal, physical work as a beekeeper, the winter allows the bees to get to know you in a different way. By greeting them, checking their entrance, breathing in their scent, clearing snow from the landing board, tapping on the box to hear their hum, talking to them—all of these things engage the bees on a very special level of connection with their beekeeper.

The winter months are also ideal for envisioning the next season with the bees—for encouraging them to re-queen, or to swarm, or not to swarm too much, or to stay away from the pristine lawn when the dandelions are trying to bloom, or to learn to practice good hygiene and diligently clean out the varroa mites, etc. Our thoughts are realities, and the power of setting intention is already a great help to the bees and to our practice when it comes time to be hands-on again.

In my experience this regular communication and practical encouragement to grow together in a positive direction is a great support for those helpful qualities to manifest and for us to continue to encourage them through our careful attention, creativity, and nimbleness year-by-year.

It is through loving devotion that the doors have opened for us to continue forward with the bees in a healthy mutual dependence. The bees will continue to show us the way if we ask and then listen carefully. What do you need? How can I serve you?

Chapter 8

Troubleshooting Situations and Common Interventions

This chapter is intended to share insight into the most common challenges that call for us to act on behalf of the bees, including how to recognize each of these situations and what might be done to help. Many beekeeping challenges from the conventional stream will not be dealt with in this chapter, as the Spikenard Sanctuary beekeeping methods prevent these from occurring in the first place (ex. supersedure queens, young queens regularly failing, queen disappearance, Colony Collapse Disorder, and the 30 or so common diseases including nosema, foulbrood, chalk brood, etc.). By allowing colonies to raise their own queens, swarm, build their own comb, eat their own honey, and live in hives that promote their warmth and scent, we have already eliminated a whole host of all-too-common issues. The need for intense treatments, chemicals, and regular interventions to prevent diseases, viruses, pest explosion, queen failure, etc., are in large part due to artificial queen rearing, sugar feeding, pollen patty feeding, the use of foundation, the prevention of swarming, and so on.

Unfortunately, there are plenty of problems that we all have to learn to deal with, no matter what our beekeeping practices are or where we live. We will each have lesser or greater problems with certain pests, challenging weather events and seasons, pesticides, toxins, pollution, EMF radiation, climate change…. We can classify these as *external challenges*—ones that come to the beehive from the world around them. And we place these in relation to *internal challenges* which arise as challenges from within the beehive itself.

As beekeepers and land stewards, we have a voice that needs to be heard in our communities when it comes to the external challenges that bees and other life forms are facing. Addressing these challenges often calls for activism through community organization and education because systemic cultural issues underlie them. As for the internal challenges that each beehive will face (sometimes due to the external challenges and sometimes not), that's where we can make a great and direct impact on the life of a colony with helpful interventions.

Troubleshooting Situations
Hive Poisoning

Spraying for mosquitoes, applying Roundup against the dandelion in suburban lawns, flyover pesticide applications of monocultures, and many other applications of pesticides, herbicides, and fungicides can be the direct cause of colonies becoming poisoned.

Observations: One day the hive is flying strongly, the next day there are hundreds and thousands of bees lying on their sides or on their backs near the entrance and around the front of the hive, dying or already dead. Often the bees' legs will be slowly groping into the air with some fully-body convulsions as they die.

Action Steps: Move the hive, that evening when all the bees are inside or early the next morning before the sun comes up, at least three miles away and prepare a new hive set up there. Make this hive cavity roughly half the size of the original (i.e. if she was sitting in two deep boxes when she was poisoned, bring her down to one). Place one or two combs of capped honey into the new hive and leave the rest of the woodenware empty or with a little comb to help guide their building. We want this colony to sweat a lot of wax to help cleanse their bodies.

Prepare a jar of Bee Tea (as outlined in Chapter 2). In the morning before much flight begins, inspect the hive, carefully looking for the queen. Once she is located, place the comb that she is on in the middle of the new hive with empty bars/frames on either side of her and one honeycomb next to each end-wall. Then gently and carefully brush and shake the remaining bees into the new box.* Once the majority are in, carefully and slowly close the lid of the hive so that the focus of orientation becomes the hive entrance. Shake or brush the rest of the bees out in front of the hive. You can set up a plywood board or a bed sheet as a ramp leading up to the entrance if you like.

153

This is quite a big event, and it will certainly be painful to remove the bees from their hive body, but it is definitely the best chance of saving a hive that has been poisoned. The poison is pervasive, so we want to give them a fresh start. As you monitor the hive in the coming days, you will start to see the number of dead bees found at the hive entrance each morning get smaller and smaller. Keep feeding her Bee Tea for a few weeks so she can focus on healing and growing her inner structure again. Do a health check in about two weeks to monitor comb building, brood in all stages, and for general signs of healthy bees diligently going about their work. If you want to move the hive back to her original location, wait three to four weeks until her new combs are solid and braced enough to the frames or hive walls to make transporting the hive safe without the risk of the combs swaying and falling.

* As you are going through and brushing the bees off of the combs, it's good to be thorough so that you are able to get every bee off of each comb. Once a comb is clear of bees, place it in a secure location (such as a box with a lid), so that bees don't come back to it but instead focus on integrating into their new hive body. You should be able to brush the bees one-by-one off of the combs and then securely store the combs away.

Aggressive Hives

Relationship is the big question when it comes to aggressive hives. In what way is the hive aggressive and under what conditions? A few notes about this:

• In the course of the season, almost all beehives become less inclined to tolerate health checks of the brood nest in the fall, while most are totally fine with this in the spring.

• In the course of the day, the hive seems to be the most comfortable between 10:00 a.m. and 2:00 p.m. through the course of the warm months. When the temperatures cool off and work needs to be done, it is generally best to wait for the warmest time of the day, even if that is 3:00 p.m. in the afternoon in the spring or fall.

• Do you only visit the hive when you have work to do? Especially with aggressive hives, practice going and sitting by the hives every day, or as many days of the week as is feasible for you. Go just to say hello, to breathe with them, to observe. Maybe place your hand by the entrance to let a few bees smell and touch you. Getting comfortable with each other is important in building trust. The more positive interactions as you can make with the beehive the better. And when you do go to open the hive, listen deeply. If you sense them getting nervous, or if they sense you getting nervous, it might be a good time to say thank you and close up the hive slowly before you create a situation that is unpleasant for both of you. When you develop rapport working together, one where the bees can let you know that it's not a good time to enter the hive and you have chosen to listen to them, then trust develops and good progress can be made.

• Have the bees gone through some challenging or traumatic event? Did the beekeeper make a handling mistake, perhaps dropping a full comb of bees or bumping the hive over with the lawnmower? Has a skunk been pestering the hive at night and eating lots of guard bees? Anything like that can raise the defenses of the beehive and make them feel less safe and potentially more sting-y.

• Genetics can play a part in aggressive hives, and a lot of conventional beekeepers would simply and unceremoniously replace the queen. This can quickly "fix the problem." But from a holistic perspective, we have created a host of other problems, including significantly disrupting the bond between the workers and the queen and the inner harmony of the colony. The effects of this disharmony are not always seen right away but will often arise a few months down the road, especially if the queen is replaced with an artificial queen that was bought.

General Advice for Aggressive Hives: Love them. Listen deeply. Take care to visit them often and develop a hands-off relationship. This will help to have a positive hands-on relationship when the time is right. And when you do go into the hive, try to work under the most ideal conditions that the season offers—for example, at 11:00 a.m. on a warm sunny day when plenty of flowers are in bloom, when you yourself are inwardly calm and outwardly prepared, when the neighbor isn't running a loud machine, etc. And remember to focus on having positive interactions as the priority rather than pushing your beekeeping agenda as the priority. This can foster great healing and help develop a deep bond between the bees and the beekeeper.

When robbing is intense early in the day, its best to leave the hive there so that the forage bees return before you move them in the evening. After closing the entrance to one-bee-wide, if the situation still continues to be intense, sometimes the best thing to do is to put a wet bedsheet over the whole hive to prevent any further robbing. This will help keep the hive cool through the day, until you can remove the sheet in the early evening before moving the hive away from harm.

Robbing Situation

This happens most often in the fall to the weakest colonies. In robbing situations, you will see bees fighting and struggling with each other at the hive entrance. You might also see bits of wax and debris littering the hive entrance area and also on the debris tray. Strong honeybee colonies, yellow jackets, hornets, and other predatory insects come to gather the honey from weaker colonies that are unable to guard their entrances. An entrance reducer helps the bees guard their hive and keep unwanted intruders out. Closing up the entrance so that there is only a small hole, just one-bee wide, will help the colony that is getting robbed to protect itself. You might have to move the hive if day-after-day the robbing is relentless. To move a hive, wait until the evening, close up the hive completely, and move it a mile away or more for at least a week. Sometimes we have also draped a wet bed sheet over the top of a colony to prevent all robbers from being able to enter if the robbing is intense.

QUEEN PROBLEMS AND QUEENLESS HIVES

"Queenrightness" is a term that is used to describe a colony where the queen is very active and strong and is laying many eggs in a beautiful brood pattern, and the rest of the colony is working in harmony with her. The queen is so essential for the life of the colony that the whole future of a colony is at risk if she is not healthy and vital. Problems with the queen can be observed in many different ways:

Spotty Brood Pattern—As the term describes, the brood nest in these situations is a bit irregular without consistent egg-laying in every cell, but with one egg here, then a few empty cells before the next one, and so on. It is especially easy to notice the spotty pattern when the pupal cells have been capped. This laying pattern means that the queen does not have a rhythmic laying pattern or is not well fertilized or has some other problem that is not allowing for a healthy brood pattern. The inconsistent laying makes it more difficult for the workers to efficiently warm and care for the developing brood. If the queen is young and is just beginning to lay, often she will settle into a nice brood pattern in the course of a couple of weeks. If the queen is many years old, it could be a sign that she is coming to the end of her

Spotty brood pattern.

productive life. Often the hive will sense this and replace her with a new queen. However, if the queen is in her prime and is still producing a spotty brood pattern, it is worth making a note of this and carefully looking at the overall health of the colony to see if the hive is able to manage the situation well or if the numbers of bees are dwindling or if there are other parallel concerns.

Only Drone Brood!—During the warm months of colony growth, you will be able to observe both worker brood and drone brood in a colony that is queenright. 10%-20% drone brood is normal. If you only see drone brood, then you have a problem with the queen. Either there is no queen and you have laying workers (see laying worker hives below) or you have an infertile queen who is not able to create workers. One of the common interventions listed below will need to be taken to help ensure that this hive has a future.

Dead queen found in front of hive.

Queenless Hives—Some common observations of queen-less hives have already been described in Chapter Three—Developing Our Sense Perception, where we discussed what a queenless hive sounds like. When you open a hive, you can also observe that the worker bees look and sound nervous and agitated in their activity. The bees are often running around frantically, and the pitch of their hum is much higher than normal. With a queenless hive, you'll observe many more bees running around the hive without perform-ing work. Observing this nervous behavior is a helpful signal to the beekeeper to make sure to check for brood in all stages.

Too Few Workers—is a condition that when first observed is not necessarily cause for alarm. From observing the en-trance you would see minimal activity, and inside the hive you might see only a small fist-sized cluster of bees attending to the brood nest. While it may indicate that the queen is not particularly productive, she may still have a well-orga-

Multiple eggs per cell—a typical observation in a laying worker hive.

nized laying pattern and efficient nest. If this is the case, then the first thing to try is tightening up the hive as much as possible (see "Tightening up," below). This situation is not too concerning in the spring but is much more so in the fall (see description of small hives getting through the winter in Ch.7). If after tightening up the hive you do not observe a consistent maintenance or growth in the number of bees, then your hive is most likely dwindling due to a weak queen and is unlikely to survive. If this is the case, we would have to respectfully kill the queen in order to perform one of the common interventions listed in the next section.

Laying-Worker Hive—In this situation you would see eggs, usually several, pasted to the sides of the cell, not nicely down at the bottom. This is the easiest way to notice the laying-worker situation. A whole group of workers begin to lay, between 5-30 in one hive. A laying-worker colony will not allow an emergency queen to be created, and they won't accept a bought queen or even a swarm. The hive itself is in the process of excarnation. The best we can do for a hive like this is to utilize the resources of this hive to the benefit of other colonies. Sometimes, all that can be saved are the combs, the honey, and the pollen. One thing you don't want to do in the laying-worker situation is nothing. A laying-worker hive begins to slowly degenerate. The entrance can become messy and their general hygiene becomes lack-luster, and bigger issues can start to enter into your apiary through these hives, like increased varroa mites, hive beetles, wax moths, disease, etc. For these reasons, it is important to dissolve or combine these colonies.

If there are still many bees, then you might want to try to incorporate them into another hive by this intervention:

• In the late afternoon, puff about 10 puffs of smoke into your laying-worker hive in order to encourage the bees to take in some extra honey. Shake the bees out in front of the colony, about 15 feet away. Immediately set another hive from a different location in the laying-worker hive position. Watch the laying-worker bees walk/fly back into the hive that you set in place. The new colony will allow the workers in and will be able to control the laying workers so that they do not become a problem. We have seen really positive situations with harmonious combinations, but it is also possible and normal for there to be some fighting at the entrance, especially at the beginning.

COMMON INTERVENTIONS FOR TROUBLESHOOTING PROBLEMS

Giving a Comb With Eggs

Giving a comb with eggs from a vibrant hive to a queenless hive is one of the easiest ways to help a queenless colony raise a new queen. The workers in the queenless hive will choose a newly hatched larva, or maybe a few, from the comb that has been given and will create emergency queens from them. An emergency queen is created by the workers from an egg that was otherwise destined to become a worker. In an emergency situation when a colony has lost its queen, the workers have the incredible capacity, through changing the form, position, and diet, to transform a worker larva into an emergency queen. These emergency queens have a much shorter life expectancy than a "true" queen who was intended to be a queen from day one (0.5-2 years for an emergency queen, 3-6 years for a "true" queen). Nevertheless, the capacity for the bees to create emergency queens gives the beekeeper the possibility of giving a queenless hive a comb with eggs so that an emergency queen can be created and the hive can become queenright.

After giving a comb with eggs, wait one to two weeks before checking to see if an emergency queen has been created. If you see the emergency queen cells developing, you can wait another two weeks before checking back in to see if the new emergency queen has been able to establish herself and begin laying eggs.

The ability to save a hive by giving a comb with eggs and allowing the bees to create an emergency queen is one important and helpful reason to have at least two of the same hive type in your apiary (ex. two Langstroth hives, rather than one Warre hive and one Langstroth hive). It's a simple procedure to offer one frame from the same hive type to another, but it's a much more challenging situation to have to cut active combs or modify woodenware in order to give a comb of eggs to a hive in need.

Making a Combination

Making a combination consists in integrating two hives together. This is done in order to preserve or transfer worker bees from a weak or dying colony to another that could benefit from having more workers and other resources. When we see a hive is not doing well and we have determined that she will likely not be able to survive long into the future, we might wonder what can be done to preserve the still-healthy aspects of the colony.

Sharing honey, pollen, and combs between hives is something that can be done without any special precaution except for checking carefully to make sure that we are not moving bees along with those. Hives will receive those gifts readily without any special treatment needed to encourage them to do so. Transferring honey from a colony that has excess to one who doesn't have enough is a common yearly task in the fall when we are preparing for winter.

When it comes to integrating the bees from one colony to another, we have to take certain measures to reduce stress and encourage a slow and peaceful integration. First, we will need to make our combination in such a way that there is only one queen who will unite the workers from both hives together. This might mean you will need to kill the queen in the weaker colony so that her workers can align themselves to the queen from the stronger colony. Removing the weak queen should be done one to three days before making the combination in order to help the queenless workers with their transition and alignment.

For a successful combination, go to the now-queenless weaker colony (as described above), and puff smoke more heavily than normal into the colony. Five to ten puffs will be enough. This will encourage the workers to fill up their crops with honey which will be a helpful housewarming gift when they arrive in their new hive body. Gather the combs that are active with bees, and any other choice combs that you'd like to be part of the combined colony, and put them all together in a box. Bring a piece of construction paper or newspaper and cut slits into it. Open the stronger colony and place the paper above the highest box (above the inner cover if you are using Langstroth). Place the box with the queenless hive above the paper and close with the lid/roof. In this way, the weaker colony will be separated from the stronger colony by the slitted paper, which will help them warm up to each other. There might be a little fighting out in front of the colony—just a few little fights and more dead bees than usual may be observed. In the course of the next one to three days, the weaker colony will integrate into the stronger. Then you need to go in, remove the paper (or what is left of it!) and remove the box that was used to integrate the weaker colony. Check back in another two weeks to look inside and make sure all is well!

Dissolving a Hive

When a colony is past the point of rescue, the beekeeper dissolves the hive to prevent sharp increases in robbing, pests, and disease which could also have detrimental effects on other hives in the apiary or the area. For the bees, dissolving a hive is the equivalent of "putting an animal down" to end the suffering. Dissolving a hive consists in shaking all of the bees out in the grass in front of the hive, comb by comb, and then removing the combs and hive body for cleaning, storage, or use in another colony.

If you are dissolving a colony that still has a queen, you will need to kill the queen before shaking the bees out.

Feeding Bee Tea as a Healing Support

Feeding for building up nucs, swarms, and packages was described in detail in Chapter 2. Emergency feeding on fondant was described in detail in Chapter 7.

Healing tea is a great support for the honeybees. In addition to building up food stores in the hives, the Bee Tea can help the bees to overcome challenges such as poisoning, nosema, foulbrood, and other issues affecting the digestive, metabolic, or immune system of the colony. For helping to address these problems and for more generally giving a colony a little boost, here is a helpful recipe:

For general strengthening – quantity for two treatments
  -Bring 3 cups of water to a boil. Take off of the stove
  and add:
  • ½ tsp. each of: chamomile, yarrow, stinging nettle,
  peppermint, and dandelion blossoms
  • ¼ tsp each of: sage, hyssop, thyme, lemon balm

  - Let steep for 10 minutes. Strain through a cloth or fine
  colander. Add another 3 cups of cold water and let cool
  until lukewarm.
  - Add 1 cup (1/2 lb.) of good honey. Stir well.

This quantity is good for two treatments or two hives. If
you only have one hive, keep the second jar in the refrigerator until used, warming up to room temperature before you
give it to the bees.

Synthesis of Feeding

Reasons for Feeding	What to Do
Not enough forage / too much competition	Feed Bee Tea Mixture; Reduce # colonies; Plant pollinator forage
For winter storage	Give honeycomb; Feed Bee Tea Mixture; Honey harvest in spring
New colony comb building	Give honeycomb; Feed Bee Tea Mixture
To give a health boost	Feed Bee Tea Mixture
Emergency winter feeding in cold weather to prevent starvation	Feed Fondant paste

Tightening Up the Hive as a Healing Support

Chapter 6 describes in detail about how to consolidate and tighten up the hives in order to prepare for the winter months. Tightening up can also help a colony more fully inhabit their hive cavity, take control of the combs and pest situation, and promote a robust increase in hive warmth and scent. The bees have the incredible capacity to create excess warmth to help burn off infection and disease, similar to a person having a fever. Tightening up the colonies is almost always our first step when we see a hive that is shrinking, dwindling, or not fully inhabiting their hive body. The simple act of making a cavity smaller for a colony can have a hugely beneficial effect and help a colony get back on their feet.

Replacing Queens and Seasonal Abundance of Queens

The healthiest and strongest queens will likely come from your own apiary if you are following the Sanctuary Beekeeping methods and allowing natural queens to develop. The seasonal abundance of queens is most notable in the spring during swarm season. As colonies prepare to swarm, they create up to a dozen new queens, all of which are viable candidates to help support a queenless hive. Bringing a comb with a developing queen cell on it to a hive in need from a hive that has excess is a very easy way to establish a new and healthy queen in a queenless hive. During spring swarm season is also the best time to share this resource of abundant queens on a local level with your fellow beekeepers.

Queens also tend to be available through late-season swarms—swarms which come later in the summer or early fall and often only contain a handful of bees along with a viable queen. Late-season swarms are often the result of natural queen replacement within a colony where new queens are raised in the late summer or fall and one of them takes over the work from the old queen. After the new queen is born, chosen, and begins to lay eggs, the old queen and the other new queens will leave the colony behind. Often a small group of workers will leave the hive with these queens and land in a little clump somewhere relatively close by. These little late-season swarms can be a great resource of queens for queenless or struggling hives in the fall.

Chapter 9

Losing a Hive

Losing hives each year is a reality that every beekeeper must face. It is often cited that in an ideal environment the natural hive losses are roughly 15% per year. At Spikenard Farm Honeybee Sanctuary the average hive loss was 12% per year in the ten-year period from 2010 – 2020. On a wider scale it is common for the conventional beekeeper to budget for 45% hive losses in the yearly business plan. Many small-scale beekeepers will have seasons where total loss of colonies occurs. Unfortunately, this is so, but I mention it here because no matter how many colonies you lose or keep, there are other beekeepers sharing your plight. Taking care of both living hives and hives that have died is part of the regular processes of beekeeping. We feel a great and important responsibility in accompanying hives towards their deaths and in caring for hives that have died. And, of course, we do all we can to prevent this from happening—the whole of Sanctuary Beekeeping is intended to serve the health and longevity of our honeybee colonies.

Accompanying a hive to its death is the complementary action to accompanying a hive through the process of swarming. There manifests before us in both instances the creative possibility of complete transformation. We honor

What kind of pattern was the brood nest?

these as sacred moments with wonder, reverence, and presence.

In the case of hives dying, we have two major activities to accomplish. First, to learn all that we can about what was going on in the inner life of the colony by carefully inspecting, observing, and taking stock of all that has been left behind. There is a great deal of information that we can gather from close inspection of the bees, brood, comb, honey, pollen, etc., that will help us to determine the possible causes of death (if we don't already know what has led to this death). These are moments that can be great opportunities to learn and to adjust our future beekeeping practices accordingly. The second major activity is to clean up the hive, including taking care of the equipment, wax, propolis, and honey. Especially wax and honey must be removed and worked with quickly, or else nature will take its course—the honey will be removed by other insects and the wax will be chewed up by wax moths.

Steps for Evaluating a Lost Hive

Keeping notes and records of each hive in its progression through the seasons is often essential to being able to accurately reflect upon a colony that has died and decipher what happened. Even simply ruling out certain possibilities will be made much easier if there are clear and consistent records to look back on.

Once you have lost a hive, it's helpful to look on the screen bottom board and see the bees that have fallen there and look carefully to see if a queen is there. This lets you know that the colony did die with a queen present. Then you want to go up and inspect the combs. Look for brood! Is there brood in all stages? Look carefully to see if there is one egg per cell. Look at the bottom of the combs: are there a whole bunch of queen cells there? Or are there emergency queen cells sitting in the middle of the comb? This lets you know that there was an emergency queen situation and that quite possibly the colony died without a queen. Look at the queen cell. If it's gnawed out on the side, it means that this developing queen was killed before she was able to hatch. Here are some other helpful lines of inquiry when inspecting a colony that has died:

What kind of pattern was the brood nest? Was it a spotty brood pattern? Can you see eggs? One egg per cell? Can you find a queen? Etc.

Look for signs that might indicate the presence or absence of a queen. Queen failure is a fairly common diagnosis for hives that have artificial/grafted queens, but it also sometimes happens in a Sanctuary setting.

How many bees were left in the hive? None? Hundreds? Thousands?

Look for signs that might indicate how many workers were in the hive, and compare this with your records through the season. An unsustainable and gradual loss of workers is often due to some underlying constitutional condition, a weak queen or a high varroa mite load, for example. Or sometimes a sudden loss of workers can be linked to a very specific event like a robbing situation or hive poisoning. These could all result in a colony losing a critical mass of worker bees required to sustain the hive.

Was there plenty of honey? Where was the honey in relation to the cluster of bees?

Look to see if they might have starved. Running out of food is quite sad. Ensuring the bees have enough honey is our central task as beekeepers. Even so, it does happen to nearly every beekeeper at least once if not many times if you live in an ecosystem where nectar sources are poor or disrupted by weather.

Did the hive get too cold or wet?

Look at the cluster of bees. Clusters that have died of cold/wet will have:

• many bees who have crawled into empty wax cells

• arrangement of bees that are all facing upwards, heads towards the top of the hive with their wings all pointed in the same direction, which is the position they gather in to deflect water when it is coming down on them

• condensation in the hive and possibly moldy combs

In these situations, pay close attention to the situation in the hive. If you have a lot of ventilation and/or poor insulation, there will be more condensation problems. If you have a small cluster with way too much extra space, the hive would have probably done better to have been tightened up.

Did the bees leave?

If you find no bees or just a very few bees in the hive, it is possible that they have absconded or left the hive due to stress or challenges within the hive. Hives will abscond for different reasons, including high pest pressure, disease, and other threats that are negatively affecting the hive. An absconding hive takes as much honey as the workers can carry in their crops and abandons everything else to go off and search for a better home. A thorough inspection of what was left behind may give clues as to why the colony absconded.

Colony Collapse Disorder (CCD)

This is often confused with absconding because the symptoms are similar. CCD can be diagnosed by the peculiar situation where the worker bees abscond or die out in the landscape all of a sudden, leaving the queen and maybe a few workers behind in the hive.

We have never experienced CCD at the Honeybee Sanctuary. Nor have any of our students or community experienced CCD once they began working with natural queens, swarms, and the rest of the Sanctuary Beekeeping methods.

Did the hive succumb to varroa mites?

Look for signs of secondary infections—such as many bees who are found dead with deformed wings. In the diagnosis after your colony has died, one of the signs that show a lot of mite pressure is that in the winter months you'll find very few workers. You will probably also find white feces from the varroa mite inside some cells. It is always possible that the colony died or was dwindling due to other reasons, and the mites proliferated as the colony grew weaker and was unable to deal with them. Records from mite counting on the debris tray are very helpful in seeing the history through the season and if varroa mites contributed to the colony's death.

The feces of the varroa mite are white and can be found in the cells upon inspecting the combs.

What do We do With the Beautiful Substances and Materials That The Bees Have Left Behind?

Strive to clean up a hive as soon as possible once you have dissolved them or discovered that they have died. The honey, the propolis, and the wax are great gifts that we can save and use later, but they will not last forever just sitting in the dead beehive. If we would leave a dead hive sitting for months without cleaning it up, we would certainly come back to discover that the honey has been robbed out or fouled up by mice and hive beetles and that the combs and woodenware are full of wax moths and their larvae and pupae. Instead, clean and take care of the gifts the bees have left us as soon as possible!

Here are brief indications and ideas for what to do with these substances:

For the woodenware, consider if the boxes need to be sanded and re-stained or painted. You can use propolis tincture on the inside of the woodenware to disinfect it and then put the woodenware outside into the warm sun for a few days. If you have had any sort of serious disease such as nosema or foulbrood, use a torch to lightly burn the inside of the woodenware so as to eliminate any remaining challenges.

For the propolis, use it for both human health and bee health. Propolis is antiviral, antifungal, and antibacterial with wonderful trace minerals and healing qualities. You can use a 40% alcohol solution to make a tincture which can be taken internally for sore throat, cough, etc., or be rubbed on with a rag to the inside of the woodenware as a gift for the next colony.

For the wax comb, save the light yellow combs if you need some extras for next year. Cut out the dark combs to melt down in the wax melter. Leave a one-inch strip of straight wax at the top to help guide the next bees to build their comb along a straight template. You can save these one-inch combs and the nice yellow/white combs on a storage rack outside. We have already shared about saving and reusing wax combs, including how to store them on a comb rack to prevent wax moths (See page 87).

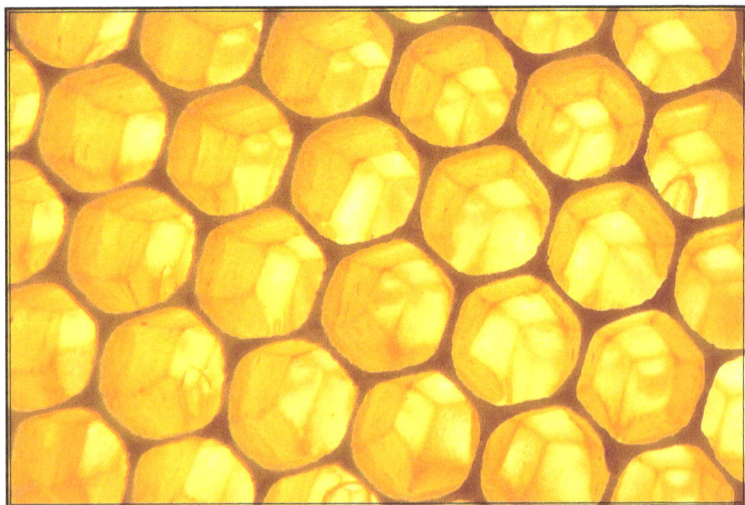

For the honey, save some of it in the combs to share with other colonies that need it through the winter or to help new swarms get started in the spring. This capped honeycomb needs to be saved inside on a comb rack (See page 86). If you have enough stored in combs, then you can extract the honey for feeding the bee, either for colonies that need help through the winter or for new swarms in the spring. At the Honeybee Sanctuary we receive whatever surplus honey is left over when the dandelion comes into bloom in April as honey for the beekeepers, friends, and community.

Afterword

For many years I have been asked: "Is there a basic biodynamic beekeeping manual where I can study and reference the information that you are teaching at Spikenard Farm?" I am very pleased that from now on, I'll be able to answer that question positively!

I hope that the students who have been yearning for a detailed printed presentation to accompany their beekeeping journey will be satisfied with what has been presented in this book and will use this as a firm ground to build their practice. I trust, through my own experience, that these beekeeping methods will prove to be a successful practical support for the health, vitality, and longevity of the honeybees.

At Spikenard Farm Honeybee Sanctuary, we walk the talk of these beekeeping principles and practices every day. As we reach the 100th year anniversary of the Bee lectures given by Rudolf Steiner in 1923—the starting point of biodynamic beekeeping—I share with you my heartfelt and humble feeling that I stand on the shoulders of giants in this presentation of Sanctuary Beekeeping. I am deeply grateful for the work of my teacher, co-worker, and friend Gunther Hauk who has cultivated and passed this legacy of beekeeping to us that approaches the bees with awe, reverence, and gratitude. He encourages us to keep an open mind, and to listen deeply, so that we might learn something new every time we go to work with the bees.

It is in this spirit that we strive forward into the future where new experiences and challenges hold the opportunity to spur us on to new insights and life-giving activity. Let us regard this beekeeping method as not the "only way" nor as just a list of recipes but as a fully inclusive transformation in the way we relate to our Earth. I waggle dance with enthusiasm and invite you to fly with us in this creative and wholesome direction towards heartening and enlightening the human-bee partnership.

Notes

Notes

www.ingramcontent.com/pod-product-compliance
Lightning Source LLC
Chambersburg PA
CBHW041313210326
41599CB00008B/257